Quantum Mechanics in the Single Photon Laboratory

Quantum Mechanics in the Single Photon Laboratory

Muhammad Hamza Waseem, Faizan-e-Ilahi and Muhammad Sabieh Anwar
Department of Physics, Syed Babar Ali School of Science and Engineering, Lahore University of Management Sciences (LUMS), Lahore, Pakistan

IOP Publishing, Bristol, UK

Supplementary material are available for this book at http://iopscience.iop.org/book/978-0-7503-3603-3.

ISBN 978-0-7503-3063-3 (ebook)
ISBN 978-0-7503-3061-9 (print)
ISBN 978-0-7503-3064-0 (myPrint)
ISBN 978-0-7503-3062-6 (mobi)

DOI 10.1088/978-0-7503-3063-3

Version: 20200701

IOP ebooks

British Library Cataloguing-in-Publication Data: A catalogue record for this book is available from the British Library.

Published by IOP Publishing, wholly owned by The Institute of Physics, London

IOP Publishing, Temple Circus, Temple Way, Bristol, BS1 6HG, UK

US Office: IOP Publishing, Inc., 190 North Independence Mall West, Suite 601, Philadelphia, PA 19106, USA

Dedicated to those who, while living in the face of danger themselves,
stand for the distressed in happenings of the global pandemic …

اگر خواهی حیات اندر خطر زی

غزالی با غزالی درد دل گفت   ازین پس در حرم گیرم کنامی
بصحرا صیدبندان در کمین اند   بکام آهوان صبحی نه شامی
امان از فتنهٔ صیاد خواهم
دلی ز اندیشه ها آزاد خواهم
رفیقش گفت اے یار خردمند   اگر خواهی حیات اندر خطر زی
دمادم خویشتن را بر فسان زن   ز تیغ پاک گوهر تیزتر زی
خطر تاب و توان را امتحان است
عیار ممکنات جسم و جان است

(علامه محمد اقبالؒ)

LIVE DANGEROUSLY
Allama Iqbal
Said one gazelle to another, ‘I will
Take shelter in the harem from now on;
For there are hunters at large in the wild,
And there is no peace here for a gazelle.
From fear of hunters I want to be free.
O how I long for some security.’
His friend replied, ‘Live dangerously, my
Wise friend, if it is life you truly seek.
Like a sword of fine mettle hurl yourself
Upon the whetting-stone; stay sharp thereby.
For danger brings out what is best in you:
It is the touchstone of all that is true.’
Translation by M Hadi Husain

Contents

Preface

For many years I had dreamed of making a laboratory for our students that could demonstrate fundamental facets of quantum physics, quantum information and quantum computing. Though our students at PhysLab, Lahore University of Management Sciences (LUMS)'s Syed Babar Ali School of Science and Engineering work with SQUIDs, Franck–Hertz tubes and lasers, I was looking for ideas that were more 'grainy', counter-intuitive and quantitative and that could directly relate to quantum interference, entanglement, density matrices, nonlocality and reveal the eerie aspects of quantum physics.

Luckily, I came across two students, Hamza Waseem and Faizan-e-Ilahi who, being students of electrical engineering in University of Engineering and Technology (UET), Lahore, converted their deep sense of deprivation of a formal education in physics to a craving for experimenting with single photons. I completed my electrical engineering from UET and so I have a fondness with this very institution. Hamza is now pursuing a doctorate in Physics from the University of Oxford (where I studied fifteen years ago), while Faizan continues his Masters in physics at LUMS.

After a year and a half of dedicated effort, struggling between coursework at their parent institution and the tough challenges we had laid out for ourselves, Hamza and Faizan managed to complete a suite of fundamental experiments on quantum mechanics. This book is about these experiments. It not only describes the experiments, but provides all the necessary working knowledge of quantum mechanics, quantum states and quantum operators that is required to motivate the reader towards performing these experiments, and providing the necessary tools to interpret and understand the experimental outcomes.

The book starts with a survey of how research labs around the world have helped create a portfolio of some wonderful experiments that can be easily translated for instructional purposes in physics teaching laboratories worldwide. As such, these experiments are a true distillation of cutting-edge research in quantum optics and quantum information science with single photons and epitomize how research in the laboratory can directly enrich the teaching of a counter-intuitive mathematical and physical framework that we call the quantum theory. Since our working units for qubits are photon's polarization state, we make a brief digression into the classical picture of light. Some of the mathematical tools introduced in the second chapter, carry over, in a quantum garb, to the quantum experiments. We also describe some experiments with lasers that can act as a spring board for the quantum experiments.

Enter the quantum description of nature with states of qubits living on the Bloch sphere. The third chapter builds just enough of the mathematical description that is required to understand how we explain quantum states, how they change with time or when acted upon by certain transforming elements, and finally, how do we measure them. Connections with photon polarization are built as we move along. This chapter also provides us the opportunity to introduce quantum entanglement, which is hailed as the quantum world's most closely guarded and most cherished secret.

The background material which spans over three chapters is not a substitute to a standard semester long course on quantum mechanics, but does provide enough material to the non-physics major, or engineer, or even the general science student, to understand how quantum mechanics unfolds in reality, inside a laboratory and is not a mere textbook thought experiment but can reveal itself in full glory with simple, 'but not simpler' instruments and gadgets.

Finally, chapters 4–6 describe the experiments. We have made a conscious effort in classifying these experiments into three categories: those that deal with the 'quantum-ness' of these quantum experiments, those that demonstrate how the quantum reality is neither local nor realist and finally, how we can measure the quantum state. All of these various kinds of experiments highlight one aspect or the other of quantum reality.

None of these experiments are novel. They have been tried many times by many people around the world. Several excellent laboratory manuals and pedagogical articles recount these experiments. We have benefited from all of these prior demonstrations. In particular, we have drawn upon the experiments described in Mark Beck's textbook 'Quantum Mechanics: Theory and Experiment' and these experiments are mostly an explanation of Beck's innovation. However, this book is not fashioned as a laboratory manual. A smart professor with some expertise in practical optics can easily decipher our narrative and reconstruct these experiments, but the text is not written as a laboratory recipe book for the student. The style is more like a textbook, but a textbook that describes real experiments. A close analogy is with the 'Experiments in Modern Physics' by A Melissinos and J Napolitino.

I would really like to say that the experiments outlined in this book are truly multi-disciplinary. Even though the motivation is understanding and revealing the quantum nature of fields, they truly make the experimenter feel what it looks like in a real research environment. Aligning optical beams, searching for the bull's eye shot of a laser focus, chasing nanosecond long electronic pulses on an oscilloscope screen, making electronic circuits that can change the voltage levels on high speed signals, computer programming and connecting experiment with an unusual theoretical structure, followed by data analysis.

In this book, we attempt to present our single-photon quantum experiments in a modular fashion, with one building on the other. The concluding chapter also presents some suggestions for teachers and researchers who are interested in reproducing these experiments, for possible pathways of building these activities in our advanced physics laboratories; or for offering these experiments, in a simple to complicated ladder, to students undertaking lab courses. The appendices provide some useful technical details about the basics and programming of field-programmable gate arrays, and also spells out an inventory which we employed in our work. The inventory is by no means comprehensive or an endorsement of one instrument over the other—it merely reflects what worked best for us, and in most cases, in the first go. All the inventory and datasets can also be downloaded from our website https://www.physlab.org/qmlab.

Muhammad Sabieh Anwar
Lahore, Pakistan, April 2020

Acknowledgments

By no means are we alone in this effort. This initiative is a pedagogical partnership and is inspired by several colleagues, most of them in the United States, who go to extreme lengths in making the teaching of quantum physics simple, intuitive, elegant and above all, a moment of joy. I like to thank Drs Mark Beck, Enrique Galvez, Richard Haskell, Theresa Lynn, Art Hobson, Suhail Zubairy and David Van Baak for inspiring me, through their open-source books, personal training and published articles, for carrying out the mission of using physics as an agent of change and well-being. Haskell and Lynn at the Harvey Mudd College led a laboratory immersion programme on single photons in 2017 that helped me learn many tools of the trade. Thanks to the ALPhA initiative for organizing these sessions.

After completing my formal studies, I joined Lahore University of Management Sciences (LUMS), a not-for-profit University in Pakistan, as an Assistant Professor. The objective that lay in front of us in 2007 was setting up a microcosm of science-infused teaching and research ecosystem in the country that would become a regional role model in education and discovery. I must say I was really lucky. The University provided me, and all of my colleagues, the academic freedom that allowed us to curate new laboratories, design innovative multi-disciplinary courses, seed research areas which were non-existent in the country and incite a generation of holistically trained science and engineering graduates who could 'think globally' but 'act locally'.

In this environment, I got the chance to cultivate a laboratory called the PhysLab (http://www.physlab.org), which was meant to create a national platform for research-inspired instruction in experimental physics with all its resources and blueprints made public, so that other institutions around us could also benefit from any humble work that we did. This lab showcases a tapestry of experiments in basic and advanced physics, ranging from fluids to cosmic ray muons, and pendulums to superconductors.

In this overarching framework, the single photon lab narrated in this book is our effort to share how quantum physics can unfold in the real world with demonstrable experiments yielding quantitative data which can be easily analyzed revealing insightful interpretations of quantum mechanics' postulates.

Even though there are three authors to this book, these experiments have been gradually built by many students and laboratory staff: Dr Wasif Zia, Obaidullah Khalid, Abdullah Khalid and Ramesh. The PhysLab depends on the unwavering support of a host of remarkable individuals and I cannot mention each one of them, but still I cannot miss Sohaib Shamim (late), Amrozia Shaheen, Junaid Alam, Muhammad Rizwan, Dr Umar Hassan in Rutgers University, Umar Hassan, Hassaan Majeed, Khadim Mehmood, Muhammad Shafique, Azeem Iqbal, Kaniz Amna, Dr Moeez Hassan, Dr Ammar Khan, Dr Waqas Mahmood, Muzammil Shah, Abdullah Irfan and Ali Hassan. I cannot thank enough my Department Head, Dr Muhammad Faryad for allowing me to pursue my passions and Arshad Maral, the department's secretary for facilitating me at every step. LUMS has also generously provided me with the financial support.

The sixth chapter of this book on quantum state tomography is joint work with my PhD student Ali Akbar, and certain extensions of this work will be part of his PhD dissertation. Ali, Alamdar and Obaidullah have immensely helped me in setting up the Ibn-e-Sahl Optics Lab (https://www.physlab.org/optics-lab/) which became the nursery for these single photon quantum experiments.

Finally, gratitude comes to a different level when your loved ones support you in invisible ways. This book was purely the idea of my father, Dr Saadat Anwar Siddiqi who has stood behind me in my career as a strong inspirer. My mother Shahida Bano has given me the prayers to carry on and define new targets for myself every time. My wife Hina Zulfiqar and my children, Fatimah and Khadeejah have given away parts of their lives in letting me engage in these academic activities. All the others specially like to thank the staff of the Institute of Physics Publishing, particularly John Navas and Sarah Armstrong and other members of the editorial and production teams who are invisible to us, for their timely support, suggestions and meticulous care for this project. It has been a true pleasure working with IOP.

Muhammad Sabieh Anwar
April 2020

We thank Dr Sidra Farid, Associate Professor of Electrical Engineering at University of Engineering and Technology (UET), Lahore, for her support and encouragement to pursue this project. We are grateful to Mah Para Iqbal and Zahra Tariq for their help in performing preliminary experiments and designing figures. Thanks are also due to the technical and support staff at LUMS.

Most importantly, we would like to express our immense gratitude to Dr Sabieh Anwar for letting us work in his lab and keeping us on our toes. The subject matter of this book signifies our transition from engineering to physics and is, therefore, especially close to our hearts. It has truly been a great privilege working at PhysLab.

Muhammad Hamza Waseem and **Faizan-e-Ilahi**
April 2020

Author biographies

Muhammad Hamza Waseem

Muhammad Hamza Waseem is pursuing a DPhil in Condensed Matter Physics as a Rhodes Scholar at the University of Oxford. His research revolves around quantum magnonics and hybrid quantum systems. During his undergraduate studies in electrical engineering at UET Lahore, Hamza helped develop classical and quantum optics experiments at PhysLab, Lahore University of Management Sciences (LUMS), and contributed to research on optical metasurfaces at ITU Lahore. Hamza is a staunch advocate of science outreach and public engagement and helps the Khwarizmi Science Society organize the Lahore Science Mela. He also co-founded Spectra (http://spectramagazine.org/), an online magazine aimed at training popular science writers in Pakistan. Hamza's academic interests lie in magnetism, optics and science education, but he also enjoys studying mathematics and philosophy for their own sake.

Faizan-e-Ilahi

Faizan-e-Ilahi is a graduate student of Physics at LUMS. During his undergraduate studies he helped to build the Single Photon Quantum Mechanics Laboratory at LUMS. He is currently working in magneto-optics. His areas of interest are quantum information and open quantum systems. He is inspired by the works and lectures of Leonard Susskind. Other than that he is interested in Urdu and Persian literature and likes to study political philosophies and dystopian literature.

Dr Muhammad Sabieh Anwar

Dr Muhammad Sabieh Anwar is an Associate Professor of physics and Dean at the LUMS Syed Babar Ali School of Science and Engineering, Pakistan. Ideas from his physics instructional laboratories have been replicated in about ten Pakistani universities. His laboratories and research are presented on https://www.physlab.org. He loves teaching physics with hand crafted in-class demonstrations and some of his courses on modern physics, electromagnetism, magnetism and quantum physics can be seen on YouTube. His research interests encompass spintronics, magnetism and optics. Sabieh is also the General Secretary of the Khwarizmi Science Society (https://www.khwarizmi.org) which is aimed at popularization of science at the grassroots levels in Pakistan. The Lahore Science Mela series has touched around 50 000 people over a period of six days in three years. Prior to joining LUMS in 2007, Sabieh was a post-doc in chemistry and materials science at University of California, Berkeley and a PhD student, as Rhodes Scholar, at the Oxford University. He is the recipient of the TWAS medal in physics for Pakistan in 2008 and the National Innovation Prize in 2015.

Abbreviations

APD	Avalanche photodiode
BBO	β-Barium borate
BDP	Beam displacing polarizer
BS	Beam splitter
CCU	Coincidence counting unit
CHSH	Clauser–Horne–Shimony–Holt
EPR	Einstein–Podolsky–Rosen
FPGA	Field-programmable gate array
HWP	Half wave plate
PBS	Polarizing beam splitter
PI	Polarization interferometer
QWP	Quarter wave plate
SPCM	Single photon counting module
SPDC	Spontaneous parametric downconversion

List of quantum optics experiments

IOP Publishing

Muhammad Hamza Waseem, Faizan-e-Ilahi and Muhammad Sabieh Anwar

Chapter 1

Introduction

Optics, the study of light, is arguably one of the oldest branches of natural sciences. Remarkably, with its subfields quantum optics and nano-optics, it continues to be an active area of research. On the other hand, quantum mechanics was born in the 20th century. Some label it as the most successful theory in physics ever devised as it correctly explains a large number of physical phenomena. Nevertheless, scientists had a hard time getting comfortable with quantum mechanics because it shook their classical world view as well as their intuition. According to the celebrated physicist Richard Feynman, 'the only mystery' of quantum mechanics is superposition [1], which has some really astonishing effects. Niels Bohr in the early part of the previous century declared 'anyone who is not shocked by quantum theory has not understood it'.

To understand superposition, let us consider a system with only two possible states, labeled 0 and 1. These states may correspond to a particle taking one of two possible paths. We can do some measurement and determine which path the particle is opting to travel. There is nothing unusual and surprising about this, but only in the classical scenario. Enter quantum superposition, and we get a bizarre possibility: the system can be in a superposition of both the 0 and 1 states, i.e. the particle is taking both paths at the same time. Quantum mechanics tells us that the system is in superposition as long as we are *unable* to determine which path the particle is taking. As soon as a measurement is performed, the superposition state collapses and the particle is found to take either one of the two paths. Hence, the act of measurement makes the particle choose one path. Now, there is a specific probability of finding the particle in either of the two paths but it is not possible to know in a confirmatory manner, prior to performing the measurement, which path the particle will actually take.

This bizarre result has made scientists uncomfortable ever since the inception of quantum mechanics. Even Einstein had serious reservations. In 1935, he along with Rosen and Podolsky [2], published a thought experiment and showed that quantum

doi:10.1088/978-0-7503-3063-3ch1

mechanics proposes 'non-local' behavior. Since he believed in localism, he concluded that quantum mechanics must either be wrong or incomplete. This debate however gave rise to the property of quantum entanglement.

Let us define quantum entanglement in terms of an optical setup. The non-linear process of optical spontaneous parametric down-conversion [3–6] can produce a photon pair that is entangled in terms of polarization. The arbitrary polarization of light is usually expressed in terms of two mutually independent orthogonal linear states of polarization, called the horizontal and the vertical. One example of an entangled state would be the superposition of the state in which photons 1 and 2 are both horizontally polarized, and the state where both photons are vertically polarized. Here, superposition implies the photon pair to be in the two aforementioned states simultaneously. Again, the photons in the pair will be entangled as long as no measurement to determine their state is made. If we use a horizontally or vertically oriented polarizer to determine the state of one of the two photons, the measurement collapses the entangled state and the other unmeasured photon is predicted to be in the same polarization state. For example, if photon 1 is measured to be horizontally polarized, we immediately know that photon 2 is also in horizontal polarization. Hence, measurement of the polarization state of one photon deterministically determines the polarization state of the other photon of the entangled pair. This strange result of quantum mechanics was termed 'spooky action at a distance' by Einstein.

The Einstein–Podolsky–Rosen (EPR) thought experiment, where particles are entangled in the momentum degree of freedom, remained untested until 1964, when John Bell formulated a set of experimentally testable inequalities that could prove the local or non-local nature of reality [7]. The 'Bell's inequalities', if verified, would de-bunk non-locality and prove quantum mechanics incorrect or incomplete. On the other hand, violation of these inequalities would be a verdict in favor of quantum mechanics' most bizarre prediction: non-localism. A series of experiments [8] have indeed shown the unequivocal violation of Bell's inequalities, hence proving that nature is non-local and quantum mechanics as we know, is *indeed* a valid theory.

In the last few years, technological advancement has caused noteworthy progress in investigating and harnessing the non-local properties of nature. One of the remarkable results is quantum teleportation [9], where the quantum state of a particle is 'teleported' to a remote location by exploiting entanglement. Subsequent work on this theme has given birth of the sister fields of quantum information and quantum communication [10].

The technological 'quantum leap' has also brought quantum mechanical experiments to the tabletop by a significant reduction in the size and cost of the apparatus involved. One class of such experiments comprises single photons. These experiments, although minimalistic, are powerful enough to demonstrate the fundamentals of quantum mechanics that have, in the last century, eluded many a great physicist. The same quantum mechanical principles also lie at the heart of cutting edge applications in quantum computing and cryptography [10]. The relative ease to produce pairs of entangled photons has brought us to the juncture where we can do quantum mechanics with relatively simple and affordable optical setups [6].

In particular, a number of experiments incorporating individual photons or correlated photon pairs have been devised for supplementing the undergraduate and post-graduate teaching of quantum mechanics [11, 12]. Many of these experiments have been described in the last two decades. Some of the important experiments include 'proof' of the existence of photons [3], interference of single photons [13, 14], local realism tests [6, 8, 15] and quantum erasure [16]. This is a book that describes some of these experiments.

Experiments which 'vindicate' the quantum reality of nature fall into various categories. In this book, we will use single photons as our archetypal quantum bits and focus on their polarization degree of freedom. This starting chapter surveys some of the key experiments and recounts major developments that have led to making these verifications more accessible and approachable by physics educators and students.

Apparently, many modern physics textbooks portray the photoelectric effect as an early example of departure from the classical world. This is true, but only partially. Contrary to popular belief held by many students (and teachers), explanation of the photoelectric effect does not strictly require the existence of photons [17, 18]. Lamb and Scully showed that the photoelectric effect could be explained using a semi-classical model where the detector atoms are considered quantized but the light is deemed classical [19]. An experiment to prove the existence of photons should rather prove the 'granular' nature of light. The results of such an experiment should not be explicable through the classical wave theory of light.

Some of these pioneering experiments were in fact described in the articles [20–22]. Later on, in 1986, Grangier *et al* performed a simple experiment proving the particulate nature of light [23]. They observed coincidences between photo-detections at the transmission and reflection outputs of a beam-splitter. Their results demonstrated that the field incident on the beam-splitter could be described by a single-photon state. Hence, it was experimentally shown that a single particle of light can be detected only once. In a pioneering piece of work, Thorn *et al* adapted the same experiment for an undergraduate laboratory in 2004 [3].

Experiments on interference and quantum erasure are rather closely related. When light is made to pass through an interferometer, the interference pattern visibility is dependent on the 'which-way' information that is available to the observer. If the 'which-way' information is not available or 'erased', high visibility interference fringes are obtained. For example, polarization analysis can be performed using polarizers, which can provide or erase 'which-way' information, determining the visibility of interference fringes [13, 24, 25]. Gogo *et al* [26] demonstrated a quantum eraser through correlated photon pairs. The pump laser beam is split into so-called idler and signal beams. The 'which-way' information is obtained by performing measurements on the idler beam. Other examples of quantum eraser experiments can be found in [27, 28].

Violation of local realism in quantum mechanics stems from the idea of the EPR paradox [2]. Bell, through his inequalities, pointed out that violation of local realism in quantum mechanics could be experimentally tested [7]. Bell's work motivated similar testable inequalities, which are now collectively termed as Bell's inequalities.

Almost all of these inequalities verify the foundational correctness of quantum mechanics and indicate that nature indeed violates local realism [8]. The first experimental test of a Bell's inequality, proposed by Freedman, was performed in 1972 [29]. The now-standard version of Bell's inequality (the CHSH inequality) for optical tests was proposed by Clauser *et al* [30] and proved by Aspect *et al* [31]. The acronym CHSH stands for Clauser–Horne–Shimony–Holt.

Greenberger *et al* pointed out the possibility of an 'all or nothing' test of local realism [32]. The original argument of Bell was statistical, i.e. it was based on different classical and quantum mechanical predictions of the probability of occurrence of some results. However, in Greenberger's test, classical mechanics predicts a certain event while quantum mechanics predicts an impossible event. In all of these cases, the experimental results have been demonstrated to agree with quantum mechanics [33]. In 1993, Hardy derived a version of Bell's theorem that is significantly easier to perform experimentally and comprehend [34]. The experimental tests of Hardy's local realism have also been shown to remarkably agree with quantum mechanics [35–37].

Over the last two decades, experimental tests of Freedman's, Hardy's and CHSH inequalities have been successfully performed strictly within undergraduate lab resources [6, 8, 15, 38]. Similar resources have been utilized to perform single qubit measurement [11] and two-qubit quantum state tomography [39, 40]. Table 1.1 lists some important research groups along with their published experiments suitable for exploring quantum optics and quantum information in an advanced undergraduate or graduate research laboratory.

This book emanates from a senior year capstone project, wherein we have re-created some of the aforementioned experiments and devised some additions as well. The purpose of this work was to design a series of experiments using single-photon sources and optical setups that demonstrate the fundamental principles of quantum

Table 1.1. Some research groups with significant experimental work on single-photon based quantum mechanics.

Research group	Publications
Beck's group at Reed College	Proof of existence of photons [3, 41], single-photon interference [11], tests of local realism [6, 8, 11], quantum eraser [26], entanglement witness [42], and EPR steering [43]
Galvez's group at Colgate University	Single-photon interference [13, 44], bi-photon interference [13], time-energy interference [45], quantum eraser [46], a test of local realism [47], and the Hong–Ou–Mandel interferometer [48]
Kwiat's group at University of Illinois at Urbana-Champaign	Source of polarization-entangled photons [4], tests of local realism [37, 49], and quantum state tomography [39, 40]
Lukishova's group at University of Rochester	Single-photon sources and photon anti-bunching [50]

mechanics. We think that these experiments, currently housed at PhysLab, Lahore University of Management Sciences (LUMS), Lahore (Pakistan), can be employed as a laboratory for pedagogical purposes or as a sandbox for experimental research in quantum optics, quantum information processing and quantum computing.

This book is organized as follows. Chapter 2 briefly discuss the classical view of light along with a few preparatory experiments. Chapter 3 gives a short overview of quantum theory in terms of photons. Chapter 4 furnishes general details of our single-photon lab and documents five experiments to investigate the quantum nature of light. Chapter 5 is about three experiments related to entanglement and non-locality whereas chapter 6 discusses quantum state tomography. Finally, chapter 7 concludes our work. This book is also supplied with appendices.

References

[1] Feynman R, Leighton R and Sands M 1965 *Lectures on Physics, Vol. III: Quantum Mechanics* (Reading, MA: Addison-Wesley)

[2] Einstein A, Podolsky B and Rosen N 1935 *Phys. Rev.* **47** 777

[3] Thorn J *et al* 2004 *Am. J. Phys.* **72** 1210

[4] Kwiat P G, Waks E, White A G, Appelbaum I and Eberhard P H 1999 *Phys. Rev.* A **60** R773

[5] Migdall A 1997 *J. Opt. Soc. Am.* B **14** 1093

[6] Dehlinger D and Mitchell M 2002 *Am. J. Phys.* **70** 898

[7] Bell J S 1964 *Phys. Phys. Fizika* **1** 195

[8] Dehlinger D and Mitchell M 2002 *Am. J. Phys.* **70** 903

[9] Bouwmeester D *et al* 1997 *Nature* **390** 575

[10] Bouwmeester D and Zeilinger A 2000 The physics of quantum information: basic concepts *The Physics of Quantum Information* (Berlin: Springer) pp 1–14

[11] Beck M 2012 *Quantum Mechanics: Theory and Experiment* (Oxford: Oxford University Press)

[12] Holbrow C, Galvez E and Parks M 2002 *Am. J. Phys.* **70** 260

[13] Galvez E J *et al* 2005 *Am. J. Phys.* **73** 127

[14] Galvez E J and Beck M 2007 Quantum optics experiments with single photons for undergraduate laboratories *Education and Training in Optics* (Rochester, NY: Optical Society of America) pp 1–8

[15] Carlson J, Olmstead M and Beck M 2006 *Am. J. Phys.* **74** 180

[16] Dimitrova T L and Weis A 2010 *Eur. J. Phys.* **31** 625

[17] Stanley R Q 1996 *Am. J. Phys.* **64** 839

[18] Milonni P 1997 *Am. J. Phys.* **65** 11

[19] Lamb W and Scully M 1969 *Volume in Honour of Alfred Kastler* (Paris: Presses Universitaires de France)

[20] Clauser J F 1974 *Phys. Rev.* D **9** 853

[21] Kimble H J, Dagenais M and Mandel L 1977 *Phys. Rev. Lett.* **39** 691

[22] Burnham D C and Weinberg D L 1970 *Phys. Rev. Lett.* **25** 84

[23] Grangier P, Roger G and Aspect A 1986 *Europhys. Lett.* **1** 173

[24] Schwindt P D, Kwiat P G and Englert B-G 1999 *Phys. Rev.* A **60** 4285

[25] Schneider M B and LaPuma I A 2002 *Am. J. Phys.* **70** 266

[26] Gogo A, Snyder W D and Beck M 2005 *Phys. Rev.* A **71** 052103

[27] Eberly J H, Mandel L and Wolf E 2013 *Coherence and Quantum Optics VII: Proceedings of the Seventh Rochester Conference on Coherence and Quantum Optics, held at the University of Rochester (June 7–10, 1995)* (Berlin: Springer)
[28] Barrow J D, Davies P C and Harper C L Jr 2004 *Science and Ultimate Reality: Quantum Theory, Cosmology, and Complexity* (Cambridge: Cambridge University Press)
[29] Freedman S J and Clauser J F 1972 *Phys. Rev. Lett.* **28** 938
[30] Clauser J F, Horne M A, Shimony A and Holt R A 1969 *Phys. Rev. Lett.* **23** 880
[31] Aspect A, Grangier P and Roger G 1982 *Phys. Rev. Lett.* **49** 91
[32] Greenberger D M, Horne M A and Zeilinger A 1989 Going beyond Bell's theorem *Bell's Theorem, Quantum Theory and Conceptions of the Universe* (Berlin: Springer) pp 69–72
[33] Pan J-W, Bouwmeester D, Daniell M, Weinfurter H and Zeilinger A 2000 *Nature* **403** 515
[34] Hardy L 1993 *Phys. Rev. Lett.* **71** 1665
[35] Torgerson J R, Branning D, Monken C H and Mandel L 1995 *Phys. Lett.* A **204** 323
[36] Di Giuseppe G, De Martini F and Boschi D 1997 *Phys. Rev.* A **56** 176
[37] White A G, James D F, Eberhard P H and Kwiat P G 1999 *Phys. Rev. Lett.* **83** 3103
[38] Brody J and Selton C 2018 *Am. J. Phys.* **86** 412
[39] James D F, Kwiat P G, Munro W J and White A G 2005 On the measurement of qubits *Asymptotic Theory of Quantum Statistical Inference: Selected Papers* (Singapore: World Scientific) pp 509–38
[40] Altepeter J B, Jeffrey E R and Kwiat P G 2005 *Adv. At. Mol. Opt. Phys.* **52** 105
[41] Beck M 2007 *J. Opt. Soc. Am.* B **24** 2972
[42] Beck M N and Beck M 2016 *Am. J. Phys.* **84** 87
[43] Dederick E and Beck M 2014 *Am. J. Phys.* **82** 962
[44] Galvez E, Malik M and Melius B 2007 *Phys. Rev.* A **75** 020302
[45] Castrillon J, Galvez E J, Rodriguez B A and Calderon-Losada O 2019 *Eur. J. Phys.* **40** 055401
[46] Pysher M *et al* 2005 *Phys. Rev.* A **72** 052327
[47] Gadway B, Galvez E and De Zela F 2008 *J. Phys.* B **42** 015503
[48] Carvioto-Lagos J *et al* 2012 *Eur. J. Phys.* **33** 1843
[49] Christensen B *et al* 2013 *Phys. Rev. Lett.* **111** 130406
[50] Bissell L 2011 Experimental realization of efficient, room temperature single-photon sources with definite circular and linear polarizations *PhD Thesis* University of Rochester

Muhammad Hamza Waseem, Faizan-e-Ilahi and Muhammad Sabieh Anwar

Chapter 2

Classical nature of light

Before investigating the quantum nature of light, we would like to briefly review its classical explanation. This will serve two purposes. First, it will help familiarize the theoretical and experimental tools required to formally undertake the study of light. Second, it will help us draw and appreciate an analogy between the classical description of coherent light and the quantum description of a beam of single photons. For instance, we will see that polarization analysis for an optical beam fetches us identical results for both quantum and classical descriptions.

Classical optics is based on an elegant synthesis of electricity and magnetism and sufficiently explains a great number of optical phenomena. Because of its wide scope and range of applicability, quantum optics is often not talked about in conventional undergrad courses on optics which mostly revolve around only geometrical and physical optics. Some recommended books for classical optics include [1, 2] and [3].

In this chapter, we present the classical description of light backed up by a few possible experimental investigations. These were conducted by the authors at PhysLab, LUMS, and constitute what we call the *Ibn-e-Sahl Corner for Optics*.

2.1 Electromagnetic waves

The theory of light as an electromagnetic wave was developed by James Clerk Maxwell in the 19th century and is considered by many to be one of the greatest achievements of classical physics. The study of electricity and magnetism is encapsulated in Maxwell's equations which can be stated as

$$\nabla \cdot \mathbf{D} = \rho, \tag{2.1}$$

$$\nabla \cdot \mathbf{B} = 0, \tag{2.2}$$

$$\nabla \times \mathbf{E} = -\frac{\partial \mathbf{B}}{\partial t}, \quad \text{and} \tag{2.3}$$

doi:10.1088/978-0-7503-3063-3ch2

$$\nabla \times \mathbf{H} = \mathbf{J} + \frac{\partial \mathbf{D}}{\partial t}, \tag{2.4}$$

where **E** and **B** are electric and magnetic fields, ρ is the free charge density, **J** is the free current density, $\mathbf{D} = \varepsilon_0 \varepsilon_r \mathbf{E}$ is the electric displacement and **H** is a quantity sometimes called the magnetizing force. The relationship between the magnetic field and magnetizing force is $\mathbf{B} = \mu_0 \mu_r \mathbf{H}$. The variables ε_0, ε_r, μ_o and μ_r are called the free space permittivity, relative permittivity, free space permeability and relative permeability respectively. The relative terms indicate the role of the medium through which the electromagnetic waves are propagating and are equal to unity in vacuum.

Equation (2.1) encodes Gauss's law of electrostatics. Equation (2.2) describes the equivalent of Gauss's law for magnetostatics and includes the assumption that free magnetic monopoles do not exist. Equation (2.3) incorporates Faraday's and Lenz's laws of electromagnetic induction. Equation (2.4) is a statement of Ampere's law wherein the second term on the right-hand side accounts for displacement current.

If there are no free charges or currents, we can combine equations (2.3) and (2.4) to obtain the following wave equation:

$$\nabla^2 \mathbf{E} = \mu_0 \varepsilon_0 \varepsilon_r \frac{\partial^2 \mathbf{E}}{\partial t^2}, \tag{2.5}$$

which describes electromagnetic waves with the velocity $v = 1/\sqrt{\mu_0 \varepsilon_0 \varepsilon_r}$. In free space ($\varepsilon_r = 1$), this speed is equal to $c \approx 2.998 \times 10^8\ \mathrm{m\,s^{-1}}$.

In free space, typical solutions to Maxwell's equations include transverse waves with mutually perpendicular electric and magnetic fields. To illustrate, let's consider a wave propagating in the z-direction and having an electric field along the x-axis. Taking ω to be the angular frequency of the wave, with $E_y = E_z = 0$ and $B_x = B_z = 0$, the Maxwell equations bear electric field solutions of the form

$$E_x(z, t) = E_{x0}\, e^{i(kz - \omega t + \phi)}, \tag{2.6}$$

where E_{x0} represents the amplitude of the electric field, ϕ denotes the optical phase and $k = 2\pi/\lambda$ is the wave vector. The complex form is often used to simplify mathematical manipulation. Similar solutions exist for the magnetic field and can be expressed as

$$B_y(z, t) = B_{y0}\, e^{i(kz - \omega t + \phi)}. \tag{2.7}$$

Therefore, an electromagnetic wave, such as a light wave, comprises propagating electric and magnetic fields. Waves propagate energy without moving matter. For any electromagnetic wave, we can compute this energy flow using the Poynting vector which is given by

$$\mathbf{S} = \mathbf{E} \times \mathbf{H}. \tag{2.8}$$

The Poynting vector determines the intensity (which can be described as the energy flow (power) per unit area in $\mathrm{W\,m^{-2}}$) of an electromagnetic wave. The time-averaged Poynting vector is given by [4]

$$\langle S \rangle = \frac{1}{2} c \varepsilon_0 n E_{x0}^2, \tag{2.9}$$

which shows that the electromagnetic wave intensity is directly proportional to the squared amplitude of the electric field and $n = \sqrt{\varepsilon_r}$ is called the medium's refractive index.

2.2 Polarization

An electromagnetic wave is transverse and has an electric field **E** that is perpendicular to the direction of propagation of the wave. This direction of the electric field is called its polarization. If we take an electromagnetic wave propagating in the z-direction in vacuum, this means that the electric field can be pointing at any instant in an arbitrary direction in the xy plane. Based on the formulation of equation (2.6), the total electric field can be written in general form as

$$\begin{aligned} \mathbf{E} &= E_{x0}\, e^{i(kz-\omega t)} \mathbf{e}_x + E_{y0}\, e^{i(kz-\omega t+\phi)} \mathbf{e}_y \\ &= E_0\, e^{i(kz-\omega t)} \left(\frac{E_{x0}}{E_0}\, \mathbf{e}_x + \frac{E_{y0}}{E_0} e^{i\phi}\, \mathbf{e}_y \right) \\ &= E_0 e^{i(kz-\omega t)} \left(A\, \mathbf{e}_x + B e^{i\phi}\, \mathbf{e}_y \right) = E_0\, e^{i(kz-\omega t)} \boldsymbol{\varepsilon}, \end{aligned} \tag{2.10}$$

where $\boldsymbol{\varepsilon} = (A\, \mathbf{e}_x + B e^{i\phi}\, \mathbf{e}_y)$ is a complex unit vector termed as the polarization vector and E_0 is the amplitude of the electrical field, given by $E_0 = \sqrt{E_{x0}^2 + E_{y0}^2}$. The values of the real parameters A, B and ϕ determine the polarization state of light. Due to the $e^{-i\omega t}$ factor, the polarization vector may change its orientation in space as the wave advances.

2.2.1 The polarization ellipse

In general, the polarization vector traces an elliptical trajectory, which is usually termed as the polarization ellipse, shown in figure 2.1 and provides an intuitive understanding of the polarization properties of light. We can express the ellipse using two angular parameters [5], namely the orientation angle denoted as ψ and the ellipticity angle denoted as χ. The former represents the tilt angle whereas the latter encodes the axial ratio of the ellipse. Let's use the polarization ellipse to look at some special cases of polarization.

Light is linearly polarized if the polarization vector does not change its direction as light propagates. In this case, there is a zero phase shift ($\phi = 0$) between the x and y components of the polarization vector. Hence, the polarization vector is real, given by the real coefficients A and B

$$\boldsymbol{\varepsilon} = A\, \mathbf{e}_x + B\, \mathbf{e}_y. \tag{2.11}$$

Some common examples of linear polarization include horizontal (where $A = 1$ and $B = 0$), vertical (where $A = 0$ and $B = 1$), diagonal (where $A = B = 1/\sqrt{2}$) and anti-diagonal (where $A = 1/\sqrt{2}$ and $B = -1/\sqrt{2}$).

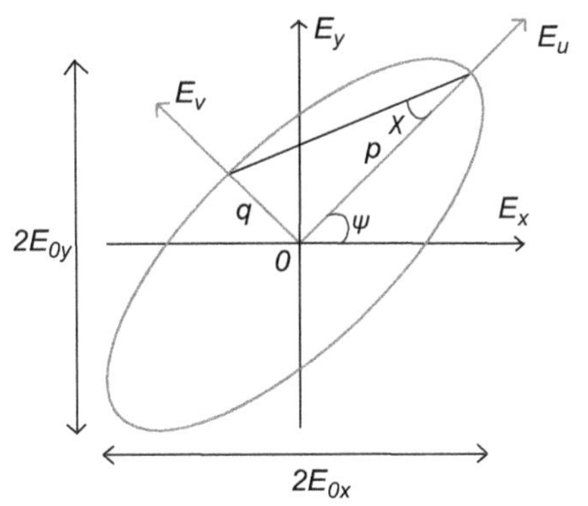

Figure 2.1. For an arbitrary polarization, the polarization ellipse can be represented in terms of the orientation angle ψ and the ellipticity angle χ. p and q denote the semi-major and semi-minor axes respectively.

If the polarization vector rotates so that it maps out a circle as the wave propagates, the light is called circularly polarized. Circularly polarized light comes in two flavors depending on clockwise or anti-clockwise rotation of the electric field vector. The polarization vector is composed of equal parts of x and y components with a phase difference of $\phi = \pm\pi/2$ between them, resulting in the polarization vector

$$\boldsymbol{\varepsilon} = \frac{1}{\sqrt{2}}(\mathbf{e}_x \pm i\, \mathbf{e}_y). \tag{2.12}$$

These linear and circular polarization states are called degenerate polarization states [5] and are illustrated in figure 2.2, which shows the locus of the tip of the electric field vector.

For an arbitrary polarization state, the polarization vector traces an ellipse. In this case, either the amplitudes of the x and y components are different or the phase difference between the two components is neither 0° nor 90°.

Light is said to be unpolarized if there are random fluctuations in the polarization vector with time. The exact parameters cannot be determined in this case. However, statistical descriptions can still be made using Stokes parameters, which are discussed in a later section.

2.2.2 Polarization manipulation

We have briefly described the commonly studied polarized states. In optical experiments, we frequently need to modify polarization. For this purpose, we use a number of optical elements. Let us briefly discuss the functional description of a few of these elements.

A *linear polarizer* transmits only that component of the polarization vector $\boldsymbol{\varepsilon}$ which is parallel to its preferred transmission axis $\mathbf{e}_\theta$. The transmitted field is given by

$$\mathbf{E}_t = E_0\, e^{i(kz-\omega t)}(\boldsymbol{\varepsilon} \cdot \mathbf{e}_\theta)\hat{e}_\theta. \tag{2.13}$$

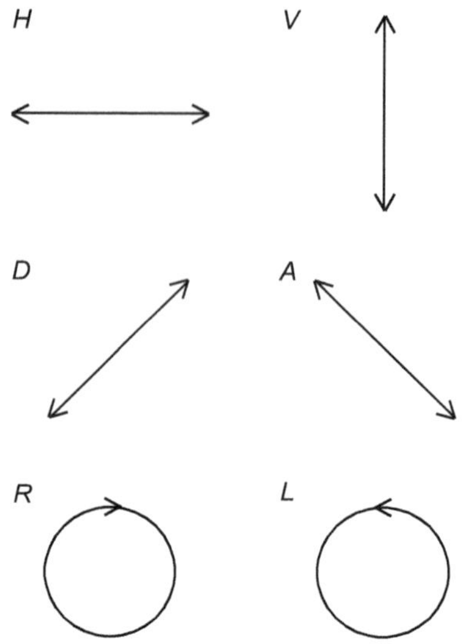

Figure 2.2. Convenient representations of the polarization vector showing degenerate polarization states: horizontal (H), vertical (V), diagonal (D), anti-diagonal (A), right circular (R) and left circular (L). The lines or circles represent the locus of the tip of the electric field vector viewed into the direction of propagation, as the wave progresses.

It is evident that the transmitted field is linearly polarized along $\hat{e}_\theta$ and its amplitude is downscaled by $|\boldsymbol{\varepsilon} \cdot \mathbf{e}_\theta|$.

Many optical elements are based on birefringent crystals. These crystals separate an optical beam having an arbitrary polarization into two beams having orthogonal polarizations. These daughter beams are termed the ordinary ray and the extraordinary ray. These two rays experience different refractive indices that could be labeled as n_o and n_e, respectively. The refractive index controls the phase velocity of waves. If these indices are different the different components of light accrue a phase difference ϕ as they propagate through the crystal resulting in a modified polarization state. A *wave plate* precisely achieves this effect, modifying the relative phase shift between the orthogonal components of the polarization vector. Quarter-wave plates (QWPs) and half-wave plates (HWPs) are the most commonly used wave plates. Depending on its orientation and the input polarization, a QWP can turn a linear polarization into an elliptical (circular) polarization or vice versa. Similarly, a HWP can rotate a linear polarization, and the rotation is controlled by the orientation of the HWP with respect to the input polarization.

A *polarizing beam splitter* (PBS) resolves the incident light into its orthogonal polarization components. In the kind of PBS that we will mostly use, the horizontal and vertical components emerge at right angles to each other.

Using a laser and the polarization modifying components we have described so far, we can perform many experiments by generating arbitrary polarization states, manipulating them and detecting the resultant light beams. To mathematically analyze such experiments in the optics of polarization, Jones calculus proves to be a valuable tool and is now described.

Table 2.1. Jones vectors corresponding to canonical polarization states. The angles α, 45° and −45° are with respect to the horizontal axis of the lab frame of reference.

Polarization	Symbol	Jones vector
Linear at angle α	$\boldsymbol{\varepsilon}_\alpha$	$\begin{pmatrix}\cos\alpha\\ \sin\alpha\end{pmatrix}$
Horizontal (linear along x-axis)	$\boldsymbol{\varepsilon}_H$	$\begin{pmatrix}1\\0\end{pmatrix}$
Vertical (linear along y-axis)	$\boldsymbol{\varepsilon}_V$	$\begin{pmatrix}0\\1\end{pmatrix}$
Diagonal (linear at 45°)	$\boldsymbol{\varepsilon}_D$	$\frac{1}{\sqrt{2}}\begin{pmatrix}1\\1\end{pmatrix}$
Anti-diagonal (linear at −45°)	$\boldsymbol{\varepsilon}_A$	$\frac{1}{\sqrt{2}}\begin{pmatrix}1\\-1\end{pmatrix}$
Left circular	$\boldsymbol{\varepsilon}_L$	$\frac{1}{\sqrt{2}}\begin{pmatrix}1\\i\end{pmatrix}$
Right circular	$\boldsymbol{\varepsilon}_R$	$\frac{1}{\sqrt{2}}\begin{pmatrix}1\\-i\end{pmatrix}$
Arbitrary	$\boldsymbol{\varepsilon}$	$\begin{pmatrix}A\\ B\,e^{i\phi}\end{pmatrix}$

2.2.3 Jones calculus

Jones calculus maps any pure polarization state to a 2×1 column vector and maps the polarization manipulating elements to 2×2 matrices. Consider again the arbitrary polarization vector

$$\boldsymbol{\varepsilon} = A\,\mathbf{e}_x + Be^{i\phi}\,\mathbf{e}_y. \tag{2.14}$$

Since we have a normalized polarization vector, the real parameters A and B satisfy $A^2 + B^2 = 1$ and ϕ is the phase difference between the orthogonal components. Conventionally, the horizontal and vertical components are taken to be $\mathbf{e}_H = \mathbf{e}_x$ and $\mathbf{e}_V = \mathbf{e}_y$. These vectors are then used as basis vectors

$$\mathbf{e}_H \equiv \begin{pmatrix}1\\0\end{pmatrix} \text{ and } \mathbf{e}_V \equiv \begin{pmatrix}0\\1\end{pmatrix} \tag{2.15}$$

to describe any arbitrary polarization vector as follows:

$$\boldsymbol{\varepsilon} = A\,\mathbf{e}_H + Be^{i\phi}\,\mathbf{e}_V = \begin{pmatrix}A\\ B\,e^{i\phi}\end{pmatrix}. \tag{2.16}$$

We call $\mathbf{e}_H$, $\mathbf{e}_V$ the *canonical basis* and the representation of any polarization state as a column vector in this basis is called a *Jones vector*. Jones vectors for degenerate and elliptical polarization states are listed in table 2.1.

Earlier, we described certain optical elements that can manipulate the polarization of a wave. The mathematical counterparts of these objects are matrices which change one Jones vector into another. These polarization modifying matrices are called *Jones matrices*. Jones matrices for the commonly used optical experiments are listed in table 2.2.

To compute the effect of a polarization changing element on a particular polarization, the Jones matrix of the optical element is simply left-multiplied to the Jones vector corresponding to the original polarization state. Usually, the polarization vector is normalized before performing Jones calculus manipulations. The output of a particular polarization element is then a unit vector multiplied by a complex constant, whose phase and amplitude determine the phase and amplitude of the electric field. The unit vector itself which encapsulates the *relative* sizes of the two basis terms and the *relative* phase determines the polarization state which in turn can be represented by the polarization ellipse described earlier.

If a beam with a certain polarization state $\boldsymbol{\varepsilon}_0$ passes through a series of polarization-modifying elements with Jones matrices $\mathbf{J}_1, \mathbf{J}_2, \ldots, \mathbf{J}_n$, the resultant polarization $\boldsymbol{\varepsilon}_n$ is given by

$$\boldsymbol{\varepsilon}_n = \mathbf{J}_n \cdots \mathbf{J}_2\mathbf{J}_1\boldsymbol{\varepsilon}_0 = \mathbf{J}_{\text{eff}}\boldsymbol{\varepsilon}_0, \tag{2.17}$$

i.e. the Jones matrices representing the respective optical elements are ordered from right to left. It is obvious that the output polarization is generally dependent on the order of the polarization manipulators. Moreover, any series of optical elements maps to an effective Jones matrix $\mathbf{J}_{\text{eff}}$ which can be found by multiplying the respective Jones matrices taken in the correct order.

Jones calculus works perfectly well for polarized light but is not applicable to partially polarized or unpolarized light. In these cases, we use Stokes parameters, which (in contrast to the amplitude description in the Jones calculus) are based on the intensity representation of polarization states and can even describe unpolarized light.

2.2.4 Stokes parameters

The Stokes parameters describe the polarization of light in terms of four observables. There are a number of ways to express the Stokes parameters, one of which is [5]

Table 2.2. Jones matrices corresponding to commonly used polarization changing optical elements. The axis orientations, represented by θ, are with respect to the horizontal axis of the lab frame.

Optical element	Symbol	Jones matrix
Linear polarizer (axis at θ)	$\mathbf{J}_P(\theta)$	$\begin{pmatrix} \cos^2\theta & \sin\theta\cos\theta \\ \sin\theta\cos\theta & \sin^2\theta \end{pmatrix}$
Half-wave plate (fast axis at θ)	$\mathbf{J}_{HWP}(\theta)$	$\begin{pmatrix} \cos 2\theta & \sin 2\theta \\ \sin 2\theta & -\cos 2\theta \end{pmatrix}$
Quarter-wave plate (fast axis at θ)	$\mathbf{J}_{QWP}(\theta)$	$\begin{pmatrix} \cos^2\theta + i\sin^2\theta & (1-i)\sin\theta\cos\theta \\ (1-i)\sin\theta\cos\theta & \sin^2\theta + i\cos^2\theta \end{pmatrix}$

$$\begin{pmatrix} S_0 \\ S_1 \\ S_2 \\ S_3 \end{pmatrix} = \begin{pmatrix} I_H + I_V \\ I_D - I_A \\ I_L - I_R \\ I_H - I_V \end{pmatrix} = \begin{pmatrix} E_{0x}^2 + E_{0y}^2 \\ 2E_{0x}E_{0y}\cos\phi \\ 2E_{0x}E_{0y}\sin\phi \\ E_{0x}^2 - E_{0y}^2 \end{pmatrix}. \tag{2.18}$$

Another equivalent expression of the Stokes parameters [6] makes apparent their relation to the polarization ellipse parameters ψ and χ, as follows:

$$\mathbf{S} = \begin{pmatrix} S_1 \\ S_2 \\ S_3 \end{pmatrix} = \begin{pmatrix} \cos(2\chi)\sin(2\psi) \\ \sin(2\chi) \\ \cos(2\chi)\cos(2\psi) \end{pmatrix} \tag{2.19}$$

where $S_1^2 + S_2^2 + S_3^2 = S_0^2$, $0 \leqslant \psi \leqslant \pi$ and $-\pi/4 < \chi \leqslant \pi/4$. For completely polarized light, S_0 is 1; for partially polarized light it is less than 1 while for completely unpolarized light it is 0. The Stokes parameters can also be graphically represented on the Bloch sphere, which is shown in figure 2.3(a). Here S_1, S_2 and S_3 represent the three-dimensional coordinate axes and **S** vector represents the particular polarization state. Table 2.3 lists the Stokes parameters for canonical polarization states.

Another graphical representation called the Poincaré sphere, shown in figure 2.3(b) is obtained if the Stokes parameters are defined as

$$\begin{pmatrix} S_0 \\ S_1 \\ S_2 \\ S_3 \end{pmatrix} = \begin{pmatrix} I_H + I_V \\ I_H - I_V \\ I_D - I_A \\ I_L - I_R \end{pmatrix}. \tag{2.20}$$

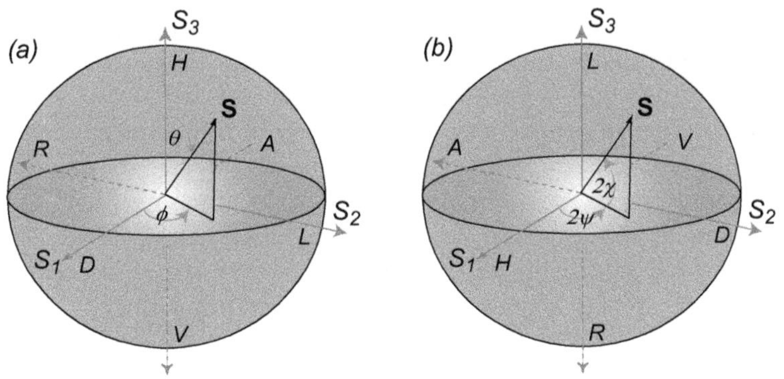

Figure 2.3. (a) The Bloch sphere representation of polarization, where the Stokes parameters S_1, S_2 and S_3 represent the three-dimensional coordinate axes and **S** vector represents the particular polarization state. (b) A sister representation of the Bloch sphere is the Poincaré sphere which maps the degenerate polarization states to different Stokes axes. The sphere shows that the Stokes parameters are related to the ellipse angles ψ and χ shown in figure 2.1.

Table 2.3. Stokes parameters for degenerate polarization states, which are perfectly polarized.

Polarization	S_0	S_1	S_2	S_3
Horizontal (H)	1	0	0	1
Vertical (V)	1	0	0	−1
Diagonal (D)	1	1	0	0
Anti-diagonal (A)	1	−1	0	0
Left circular (L)	1	0	1	0
Right circular (R)	1	0	−1	0

Comparing these definitions with those in equation (2.18), it is evident that the Poincaré and Bloch spheres are similar but map the degenerate polarization states to different Stokes axes. Although the Poincaré representation is ubiquitous in classical optics, we prefer the Bloch picture as it conforms very well to our quantum optical description of polarization.

Similar to Jones calculus, the effect of polarization changing elements on the polarization state can be computed by multiplying 4×4 Mueller matrices with the vector of equation (2.18), a system which is termed as Mueller calculus. On the Bloch sphere, these polarization manipulations are mapped as rotations and scaling of the **S** vector.

2.3 Experimental explorations

Before we embark on our expedition of setting up the experiments that demonstrate quantum mechanical effects, it is instructive to describe a few experiments within the realm of classical optics. These can help brush up many mathematical and physical concepts in optics, in addition to laying the groundwork for several useful experimental techniques on the optical bench.

As a run-up to the quantum experiments, a number of classical optics experiments can be performed. Here we describe some of the key experiments which include:

1. investigating polarization of light through Jones calculus,
2. polarization peanuts with Fourier analysis,
3. Mach–Zehnder interferometry and erasure of 'which-path' information.

We have chosen these experiments since their outcomes will be helpful in interpreting some of the quantum experiments explained in subsequent chapters. Additional classical experiments which are not directly related yet useful for training can be seen on our website https://www.physlab.org/optics-lab.

2.3.1 Light source and detection

The light source in our classical experiments is a helium–neon (He–Ne) gas laser with a wavelength of ≈633 nm and an optical power of ⩽10 mW at room temperature. The laser is unpolarized and hence polarizers and wave plates are used to generate the required polarization states.

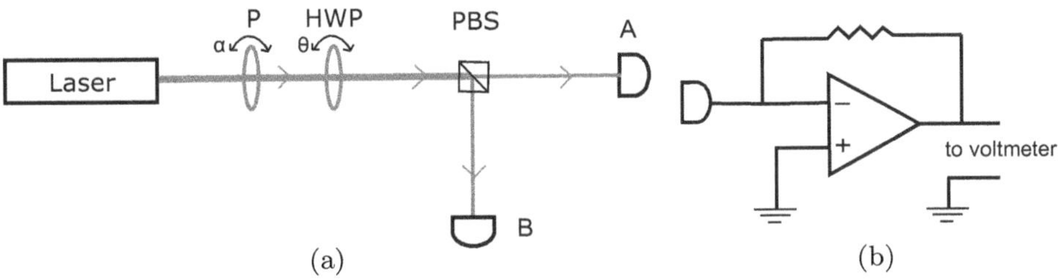

Figure 2.4. (a) Schematic diagram of an experiment to study the polarization of light and practising Jones calculus. Light from a 633 nm laser passes through a polarizer (P) oriented at α, a half-wave plate (HWP) oriented at θ and falls on a polarizing beam splitter (PBS); which lets horizontally polarized light transmit and fall on photodetector A and vertically polarized reflect towards photodetector B. (b) An IV converter converts photodetector output current into proportional voltage readable by a voltmeter.

For light detection, we use silicon photodiodes which output a current proportional to the intensity of the light beam incident on them. We convert the current to a proportional voltage through a current-to-voltage (IV) converter which is basically a trans-impedance amplifier. Using an optical detector, we do not measure the electric field associated with the optical wave. Instead, we measure the power that is directly proportional to the optical intensity I. We can define the optical intensity as the squared magnitude of the polarization vector

$$I \equiv \boldsymbol{\varepsilon}^* \cdot \boldsymbol{\varepsilon} = |\boldsymbol{\varepsilon}|^2. \tag{2.21}$$

In fact, this intensity is proportional to the magnitude of the time-averaged Poynting vector given in equation (2.9). However, the definition given in equation (2.21) will be used to compute intensity in the experiments discussed next.

2.3.2 Understanding, manipulating and measuring polarization using Jones calculus

Referring to the schematic in figure 2.4(a), we generate different linear polarization states through a polarizer and a HWP. Light from a He–Ne laser passes through a linear polarizer oriented at α with respect to the horizontal and then an HWP oriented at θ with respect to the horizontal. The polarization of light coming out of the HWP can be computed using Jones calculus and is

$$\begin{aligned} \boldsymbol{\varepsilon} &= \mathbf{J}_{HWP}(\theta)\boldsymbol{\varepsilon}_\alpha \\ &= \begin{pmatrix} \cos 2\theta & \sin 2\theta \\ \sin 2\theta & -\cos 2\theta \end{pmatrix} \begin{pmatrix} \cos\alpha \\ \sin\alpha \end{pmatrix} \\ &= \begin{pmatrix} \cos(2\theta - \alpha) \\ \sin(2\theta - \alpha) \end{pmatrix}. \end{aligned} \tag{2.22}$$

Subsequently, the light is incident on a PBS, which separates the two orthogonal components of the field, which are then detected as intensities impingent on the two photodiodes labeled A and B. These intensities are given by $I_A = \cos^2(2\theta - \alpha)$ and $I_B = \sin^2(2\theta - \alpha)$ respectively. These intensity expressions are simulated and plotted in figure 2.5.

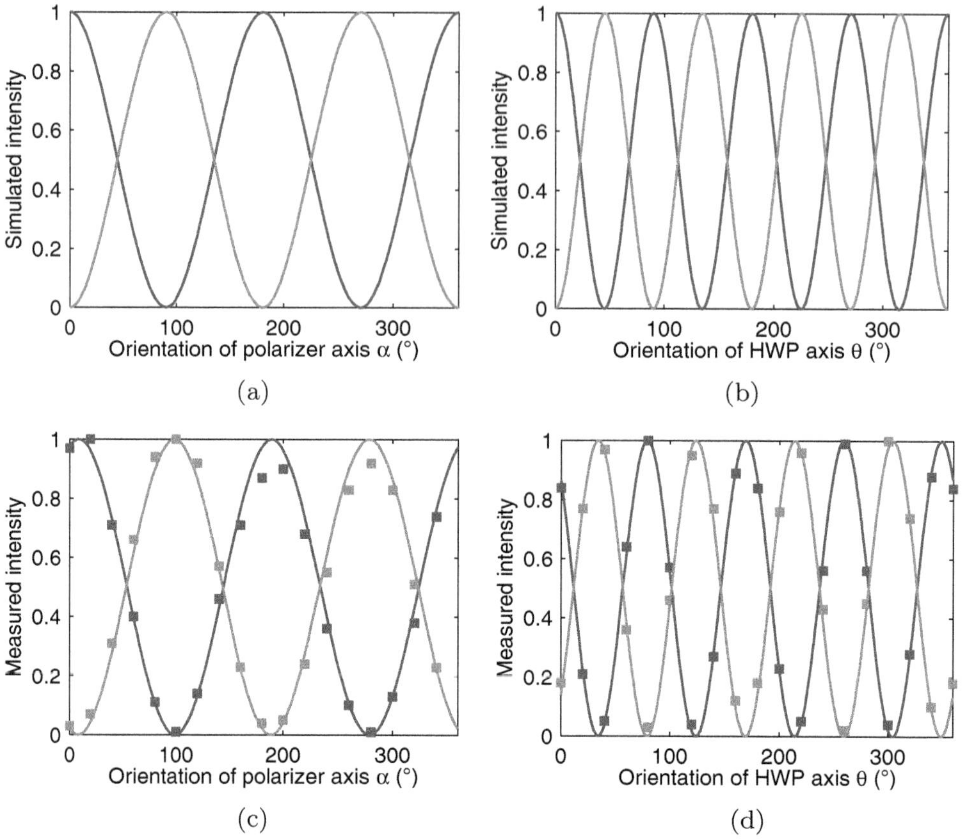

Figure 2.5. Simulated and experimental results of the experiment of figure 2.4(a). Blue plots represent detector A while red plots represent detector B. All intensities are normalized. (a) Simulated variation in intensity with respect to polarizer orientation α when the HWP is oriented at 0°. (b) Simulated variation in intensity with respect to HWP orientation θ when the polarizer is oriented at 0°. (c) Measured variation in intensity with respect to polarizer orientation. (d) Measured variation in intensity with respect to HWP orientation. The solid squares represent the measurements while the smooth plots represent the curve fits.

Each photodiode is connected to an IV converter (schematically shown in figure 2.4(b)) which converts the photodiode output current, which is proportional to light intensity, into a voltage which is then measured through a voltmeter. The measured intensities plotted in figure 2.5 show excellent agreement with curve fits of intensity expressions obtained from Jones calculus. As shown in the simulated curves, the two intensities are orthogonal to each other, confirming the basic function of the PBS. Moreover, turning the HWP shows a periodicity of $\pi/2$ while turning the polarizer shows a periodicity of π. Both of these behaviors are corroborated in the simulated and experimentally achieved results.

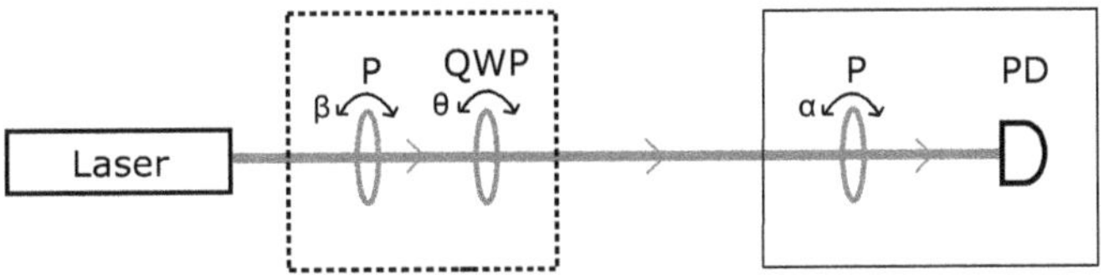

Figure 2.6. Schematic diagram of an experiment to study the polarization of light via Fourier analysis and Stokes parameters. Light from a 633 nm laser passes through a polarizer (P) oriented at β, a quarter-wave plate (QWP) oriented at θ, another polarizer oriented at α and finally falls on a photodetector (PD). The dotted box represents polarization state generator while the solid box represents polarization state analyzer. Adapted with permission from [8] © The Optical Society.

2.3.3 Fourier analysis and peanut plots

We can perform another interesting experiment to analyze the polarization state of light. This method is based on Fourier analysis and involves the Stokes parametric description of polarization. As shown in figure 2.6, a polarizer oriented at β and a QWP oriented at θ are used to generate an arbitrarily polarized laser beam. Another polarizer oriented at α is then used as an analyzer. All angular orientations are quoted with respect to the horizontal axis of the lab frame. The electric field prior to hitting the photodetector can be calculated in the usual fashion through Jones calculus:

$$\begin{aligned}\boldsymbol{\varepsilon} &= \mathbf{J}_P(\alpha)\mathbf{J}_{QWP}(\theta)\boldsymbol{\varepsilon}_\beta \\ &= \begin{pmatrix} \cos^2\alpha & \sin\alpha\cos\alpha \\ \sin\alpha\cos\alpha & \sin^2\alpha \end{pmatrix} \\ &\quad \begin{pmatrix} \cos^2\theta + i\sin^2\theta & (1-i)\sin\theta\cos\theta \\ (1-i)\sin\theta\cos\theta & \sin^2\theta + i\cos^2\theta \end{pmatrix}\begin{pmatrix} \cos\beta \\ \sin\beta \end{pmatrix} \\ &= \begin{pmatrix} \frac{1+i}{2}\cos\alpha[\cos(\alpha-\beta) - i\cos(\alpha+\beta-2\theta)] \\ \frac{1-i}{2}\sin\alpha[i\cos(\alpha-\beta) + \cos(\alpha+\beta-2\theta)] \end{pmatrix}.\end{aligned} \tag{2.23}$$

Finally, the intensity is calculated as

$$\begin{aligned} I &= \boldsymbol{\varepsilon}^* \cdot \boldsymbol{\varepsilon} = \frac{1}{4}(2 + \cos(2(\alpha-\beta)) + \cos(2(\alpha+\beta-2\theta))) \\ &= \frac{1}{2} + \frac{1}{2}\cos(2\theta - 2\beta)\cos(2\theta)\cos(2\alpha) + \frac{1}{2}\cos(2\theta-2\beta)\sin(2\theta) \\ &\quad \sin(2\alpha), \end{aligned} \tag{2.24}$$

which is a Fourier series in α of the form [7]

$$I = \frac{S_0}{2} + \frac{S_3}{2}\cos(2\alpha) + \frac{S_1}{2}\sin(2\alpha). \tag{2.25}$$

We can show that S_0, S_3 and S_1 in equation (2.25) are in fact the respective Stokes parameters of the light leaving the QWP. Consider again the polarization generated by the polarization generator

$$\mathbf{J}_{QWP}(\theta)\boldsymbol{\varepsilon}_\beta = \begin{pmatrix} \cos^2\theta + i\sin^2\theta & (1-i)\sin\theta\cos\theta \\ (1-i)\sin\theta\cos\theta & \sin^2\theta + i\cos^2\theta \end{pmatrix}\begin{pmatrix} \cos\beta \\ \sin\beta \end{pmatrix}$$
$$= \begin{pmatrix} \frac{1+i}{2}(\cos\beta - i\cos(\beta - 2\theta)) \\ \frac{1+i}{2}(\sin\beta + i\sin(\beta - 2\theta)) \end{pmatrix}. \tag{2.26}$$

Then S_0, S_3 and S_1 can be calculated from the components of this vector:

$$\begin{aligned} S_0 = I_H + I_V &= E_{0x}^2 + E_{0y}^2 \\ &= \left| \frac{1+i}{2}(\cos\beta - i\cos(\beta - 2\theta)) \right|^2 \\ &\quad + \left| \frac{1+i}{2}(\sin\beta + i\sin(\beta - 2\theta)) \right|^2 \\ &= 1, \end{aligned} \tag{2.27}$$

$$\begin{aligned} S_3 = I_H - I_V &= E_{0x}^2 - E_{0y}^2 \\ &= \left| \frac{1+i}{2}(\cos\beta - i\cos(\beta - 2\theta)) \right|^2 \\ &\quad - \left| \frac{1+i}{2}(\sin\beta + i\sin(\beta - 2\theta)) \right|^2 \\ &= \cos(2\theta - 2\beta)\cos(2\theta), \text{ and} \end{aligned} \tag{2.28}$$

$$\begin{aligned} S_1 = I_D - I_A = |E_D|^2 - |E_A|^2 &= \left| \frac{E_{0x} + E_{0y}}{\sqrt{2}} \right|^2 - \left| \frac{E_{0x} - E_{0y}}{\sqrt{2}} \right|^2 \\ &= \cos(2\theta - 2\beta)\sin(2\theta), \end{aligned} \tag{2.29}$$

which are consistent with equations (2.24) and (2.25) (and, comparing with equation (2.19), also imply that $\theta = \psi$ and $\theta - \beta = \chi$). Therefore, if for a given polarization state, we have intensity measurements at a number of analyzer orientations, we can make a Fourier curve fit to find the three Stokes parameters. This method however does not fetch us S_2 which provides information about the handedness of the polarization state. Hence, the analyzer cannot differentiate, for example, between left circularly polarized and right circularly polarized light. To determine S_2, at least one more QWP needs to be added to the analyzer.

In our warm-up experiments, we use the polarization state generator described above to generate a number of polarization states. For each state, the analyzing polarizer is oriented in steps of 10° from 0° through 360° and the photodiode output is measured. The aforementioned technique is used to determine the respective Stokes parameters, and the experimental results are summarized in table 2.4. The measured results show good agreement with theoretical predictions.

We now describe an alternative technique [8] to compute the Stokes parameters in the same experiment whose layout is sketched in figure 2.6. This technique is actually adapted from antenna polarimetry and will be revisited when we discuss the quantum experiments. Suppose that for an arbitrary polarization we have obtained the intensity profile with respect to analyzer orientations as described above. Now if we make a plot between $\sqrt{I(\alpha)}\cos\alpha$ and $\sqrt{I(\alpha)}\sin\alpha$, we get a peanut-shaped curve which geometers call a 'hippopede' and is shown in figure 2.7.

From the figure, it is evident that the polarization ellipse and the hippopede have the same orientation ψ and ellipticity χ angles. Therefore ψ can be directly measured from the tilt of the hippopede while χ can be determined by measuring the axial ratio (q/p) and subsequently using the formula $\chi = \pm\tan^{-1}(q/p)$. The Stokes parameters can then be computed using equation (2.19). Note the sign ambiguity of χ and resultantly of S_2, which implies that once again, this experimental scheme does not tell us about the polarization handedness. Table 2.4 enlists the Stokes parameters for

Table 2.4. Results of optical polarimetry of degenerate polarization states, showing a comparison between theoretical predictions and experimental outcomes. Our technique does not allow us to measure the parameter S_2. N.A. stands for 'not available'.

Polarization state	Fourier fit	Hippopede method	Prediction
Horizontal (H)	$S_0 = 1.00 \pm 0.02$	1.00 ± 0.00	1
	$S_3 = 0.97 \pm 0.02$	1.00 ± 0.00	1
	$S_1 = 0.00 \pm 0.02$	0.00 ± 0.02	0
Vertical (V)	$S_0 = 1.00 \pm 0.02$	1.00 ± 0.00	1
	$S_3 = -0.97 \pm 0.02$	-0.99 ± 0.00	−1
	$S_1 = 0.10 \pm 0.02$	0.07 ± 0.02	0
Diagonal (D)	$S_0 = 1.00 \pm 0.04$	1.00 ± 0.00	1
	$S_3 = 0.04 \pm 0.04$	0.04 ± 0.02	0
	$S_1 = 0.94 \pm 0.04$	0.99 ± 0.00	1
Anti-diagonal (A)	$S_0 = 1.00 \pm 0.02$	1.00 ± 0.00	1
	$S_3 = 0.00 \pm 0.02$	0.00 ± 0.02	0
	$S_1 = -0.98 \pm 0.02$	-1.00 ± 0.00	−1
Left circular (L)	$S_0 = 1.00 \pm 0.01$	1.00 ± 0.00	1
	$S_3 = 0.07 \pm 0.01$	0.00 ± 0.00	0
	$S_1 = -0.01 \pm 0.01$	0.05 ± 0.02	0
	$S_2 =$ N.A.	N.A.	1
Right circular (R)	$S_0 = 1.00 \pm 0.02$	1.00 ± 0.00	1
	$S_3 = -0.02 \pm 0.02$	0.00 ± 0.00	0
	$S_1 = 0.04 \pm 0.02$	0.02 ± 0.02	0
	$S_2 =$ N.A.	N.A.	−1

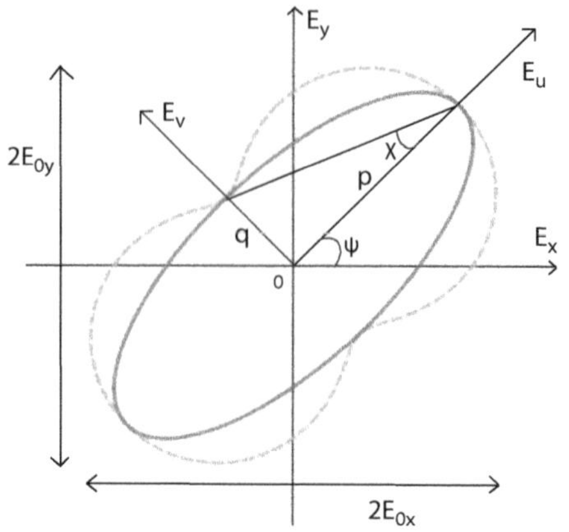

Figure 2.7. Relation of the polarization ellipse and the hippopede for an arbitrary polarization. Angular parameters of the ellipse include the orientation angle ψ and the ellipticity angle χ. p and q denote the semi-major and semi-minor axes respectively. Reprinted with permission from [8] © The Optical Society.

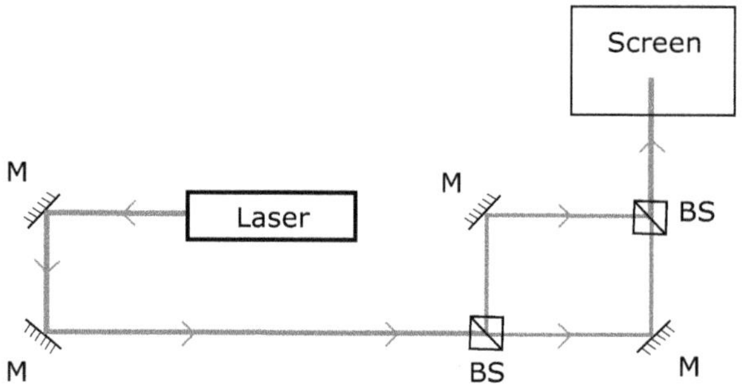

Figure 2.8. Schematic diagram of a Mach–Zehnder interferometer experiment. Light from a 633 nm laser reflects off two mirrors (M) and enters an interferometer comprising two beam splitters (BS) and two mirrors (M). Light from the two paths of the interferometer combine and fall on a screen.

degenerate polarization states obtained using this technique. The results show good agreement with both Fourier curve fit results as well as theoretical predictions. We will revisit these peanut-shaped patterns for quantum and classical light, showing experimental data in chapter 4.

2.3.4 Interference and erasure of which-way information

A Mach–Zehnder interferometer [9], shown in figure 2.8, divides the input light beam into two beams and then re-combines them. The path difference between the two component beams results in an interference pattern. Two mirrors are used to align or 'walk' the He–Ne laser beam into a beam splitter (BS), which lets half the incoming light to pass through it and reflects the rest. Each of the two beams are then reflected off a mirror in its path. The beams combine at the second BS and fall on a white screen where interference fringes can be seen. We note that each BS is a non-polarizing beam splitter. It can be observed that if one of the two interferometer beams is blocked, no interference will be apparent.

A Mach–Zehnder interferometer is also a useful device to study not only the basic principles of polarization but also the erasure of which-way information. This erasure experiment underlies an important discussion in quantum mechanics, and harbors a profound foundational significance. Let's see how.

Consider the interferometric setup in figure 2.9 where we have added three linear polarizers: one before the light enters the interferometer and one in each of the two interferometer paths. We set the polarizer P_0 at 45° and P_1 at 0°, and record the interference pattern for two orientations of polarizer P_2: at 0° and at 90°. The outcomes achieved are shown in figure 2.10, indicating that we observe no interference when the axes of the polarizers P_1 and P_2 are perpendicular to each other, implying that light has taken only one of the two paths. In contrast, high visibility fringes are observed when the axes are mutually parallel. In such a case it is not possible to determine the path taken by light as it leaves the interferometer. From another perspective, the presence of an interference pattern shows that the light has taken both paths simultaneously.

Yet another way to state this is in terms of information. In the absence of which-way information, light takes both the available paths in the interferometer and hence interference fringes are observed. In contrast, we effectively fetch which-way information by using crossed polarizers which tag the path of light. In this case, no interference is observed, implying that light takes only one of the two paths. Which-way information and interference are mutually exclusive. A knowledge of path precludes interference!

Finally, let's study the experiment mathematically, using our favorite Jones calculus [10]. The polarization state of light entering the interferometer after P_0 can be expressed as

$$\boldsymbol{\varepsilon}_{in} = \frac{1}{\sqrt{2}}\begin{pmatrix}1\\1\end{pmatrix}, \tag{2.30}$$

while the polarization state of light leaving the interferometer and hitting the screen can be computed as

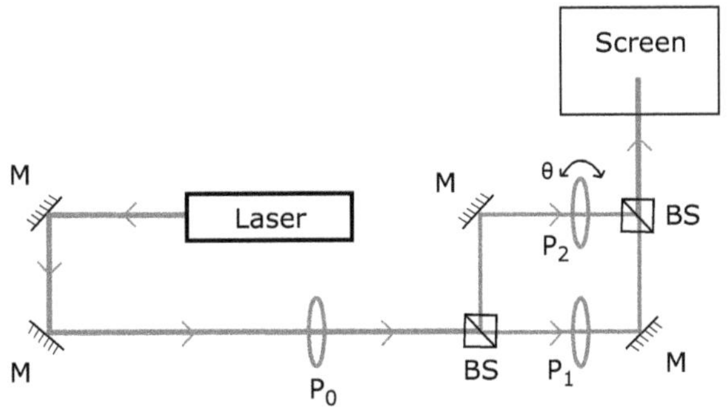

Figure 2.9. Schematic diagram of the Mach–Zehnder interferometer with three polarizers P_0, P_1 and P_2.

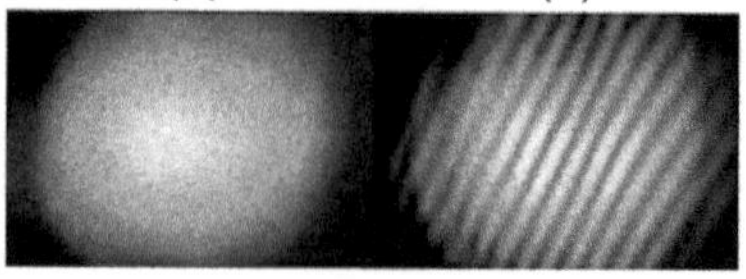

Figure 2.10. Referring to the experiment in figure 2.9, with polarizer P_0 set at 45°, (a) when polarizers P_1 and P_2 are crossed (oriented orthogonally), no interference is visible; (b) but when polarizers P_1 and P_2 are oriented with parallel axes, interference fringes are observed.

$$\boldsymbol{\varepsilon}_{out} = r\mathbf{J}_{P2}r_m r\boldsymbol{\varepsilon}_{in} + e^{i\phi}tr_m\mathbf{J}_{P1}t\boldsymbol{\varepsilon}_{in}, \tag{2.31}$$

where r and t are the reflection and transmission coefficients of the BS, r_m represents the reflection coefficient of each mirror and $e^{i\phi}$ represents the phase difference between the two beams due to small path difference within the coherence length of the laser. We take the mirror reflection coefficients r_m to be 1 and for a 50:50 BS, the coefficients r and t are each equal to 1/2. Taking P_1 to be oriented horizontally and P_2 to be oriented at θ, the output field becomes

$$\begin{aligned}\boldsymbol{\varepsilon}_{out} &= \frac{1}{2}\begin{pmatrix}\cos^2\theta & \sin\theta\cos\theta \\ \sin\theta\cos\theta & \sin^2\theta\end{pmatrix}\frac{1}{2}\frac{1}{\sqrt{2}}\begin{pmatrix}1\\1\end{pmatrix} + e^{i\phi}\frac{1}{2}\begin{pmatrix}1 & 0\\0 & 0\end{pmatrix}\frac{1}{2}\frac{1}{\sqrt{2}}\begin{pmatrix}1\\1\end{pmatrix} \\ &= \frac{1}{4\sqrt{2}}\begin{pmatrix}\cos^2\theta + \cos\theta\sin\theta + e^{i\phi} \\ \cos\theta\sin\theta + \sin^2\theta\end{pmatrix},\end{aligned} \tag{2.32}$$

for which the detectable intensity is computed to be

$$\begin{aligned}I &= \boldsymbol{\varepsilon}^*_{out}\cdot\boldsymbol{\varepsilon}_{out} \\ &= \frac{1}{32}(2 + \sin(2\theta) + \sin(2\theta)\cos(\phi) + 2\cos^2(\theta)\cos(\phi)).\end{aligned} \tag{2.33}$$

It can be seen that for P_2 set at $\theta = 0°$, I becomes $(1 + \cos\phi)/16$, which describes the presence of interference fringes. On the other hand, if P_2 is set at $\theta = 90°$, I becomes 1/16, giving a constant intensity pattern with no dependence on the path difference ϕ and hence no interference. Our observations are in agreement with the mathematical analysis. In fact, figure 2.11 shows observed intensity profiles for a number of different orientations θ of P_2.

Let us consider an interesting extension of the interferometer which is shown in figure 2.12. We place another polarizer P_3 *after* the interferometer, and setting the polarizers P_0 at 45°, P_1 at 0° and P_2 at 90°, observe the interference pattern for different orientations ψ of P_3. Mathematically, this means putting $\theta = 90°$ in equation (2.32) and multiplying the resulting Jones vector with another matrix for a polarizer oriented at ψ, as follows:

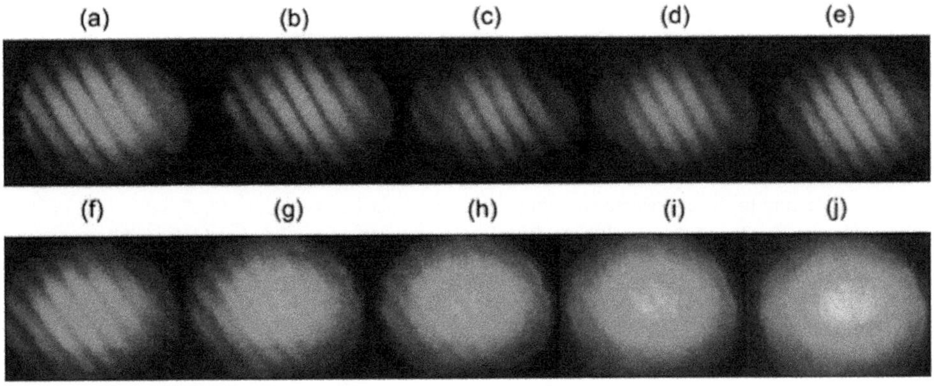

Figure 2.11. Observed intensity profiles for the experiment depicted in figure 2.9. P_0 and P_1 are oriented at 45° and 0°, respectively. P_2 is oriented at (a) 0°, (b) 10°, (c) 20°, (d) 30°, (e) 40°, (f) 50°, (g) 60°, (h) 70°, (i) 80°, and (j) 90°.

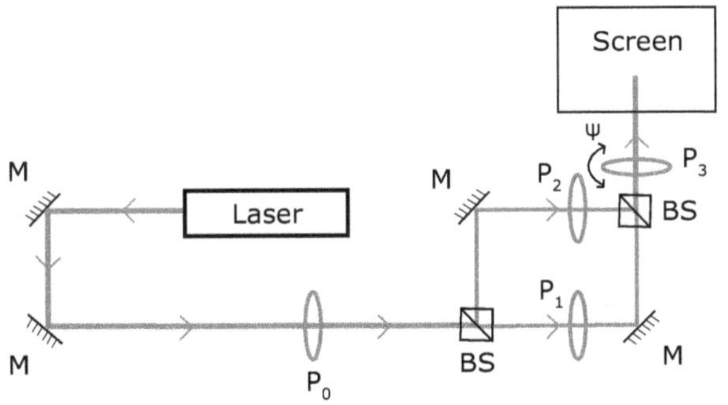

Figure 2.12. Schematic diagram of the extended Mach–Zehnder interferometer experiment with four polarizers P_0, P_1, P_2 and P_3.

$$\begin{aligned}\boldsymbol{\varepsilon}_{out} &= \begin{pmatrix} \cos^2\psi & \sin\psi\cos\psi \\ \sin\psi\cos\psi & \sin^2\psi \end{pmatrix} \frac{1}{4\sqrt{2}} \begin{pmatrix} e^{i\phi} \\ 1 \end{pmatrix} \\ &= \frac{1}{4\sqrt{2}} \begin{pmatrix} \cos^2\psi e^{i\phi} + \cos\psi\sin\psi \\ \cos\psi\sin\psi e^{i\phi} + \sin^2\psi \end{pmatrix}. \end{aligned} \tag{2.34}$$

The intensity in this case is calculated to be

$$I = \frac{1}{32}(1 + \sin(2\psi)\cos(\phi)), \tag{2.35}$$

which implies that high visibility interference fringes will be visible when $\psi = 45°$ whereas no interference will be observed for $\psi = 0°$ or $\psi = 90°$. The experimental results shown in figure 2.13 agree with this prediction.

In the setup of figure 2.9, we saw that when the polarizers P_1 and P_2 were cross-polarized (horizontal and vertical, respectively), the two paths were 'tagged' and no

Figure 2.13. Referring to the experiment in figure 2.12, with polarizer P_0 set at 45°, P_1 set at 0° and P_2 set at 90°; no interference is observed for polarizer P_3 oriented at (a) 0° or (c) 90°, but high visibility interference fringes are observed for P_3 oriented at (b) 45°.

interference could be observed. We have also concluded from the prior experiment that the absence of which-path information is necessary to obtain interference. We then added a polarizer oriented at 45° after the second BS (figure 2.12), 'erasing' the which-path information and making interference appear again.

The counter-intuitive aspect of this experiment is the delayed choice. Light, having a finite speed, hits polarizer P_3 after allegedly taking one of the two tagged paths. Just quitting the second BS, the which-path information is available and interference cannot be observed. However, as the light passes through P_3, the which-path information can be *erased* and even though we believe light originally took one of the two paths, we still see interference, implying that light actually took both paths. A process that is inserted into the apparatus after the doings with the apparatus are completed is modifying what happened earlier, the present is apparently altering the past! At least this is how the popular account of delayed choice and quantum erasure is narrated [11, 12]. We have shown instead that mathematics and Jones calculus can easily describe the experimental outcomes.

In later chapters, we will see that this phenomenon also applies to single photons, implying single-photon interference and quantum erasure! This erasure experiment is a classic example of Wheeler's delayed choice experiment [13, 14], which lies at the heart of the debate about the foundations of quantum mechanics.

Let us briefly review the fundamentals of quantum mechanics in the next chapter before beginning our experimental investigation of single photons.

References

[1] Hecht E 1998 *Optics* (Reading, MA: Addison-Wesley)
[2] Peatross J and Ware M 2011 *Physics of Light and Optics* (Provo, UT: Brigham Young University, Department of Physics)
[3] Pedrotti F L, Pedrotti L M and Pedrotti L S 2017 *Introduction to Optics* (Cambridge: Cambridge University Press)
[4] Fox M 2006 *Quantum Optics: An Introduction* (Oxford: Oxford University Press)
[5] Collett E 2005 *Field Guide to Polarization* (Bellingham, WA: SPIE)
[6] Collett E and Schaefer B 2008 *Appl. Opt.* **47** 4009
[7] Goldstein D H 2016 *Polarized Light* (Boca Raton, FL: CRC Press)
[8] Waseem M H *et al* 2019 *Appl. Opt.* **58** 8442
[9] Beck M 2012 *Quantum Mechanics: Theory and Experiment* (Oxford: Oxford University Press)

[10] Dimitrova T L and Weis A 2010 *Eur. J. Phys.* **31** 625
[11] Hobson A 1996 *Phys. Teach.* **34** 202
[12] Hillmer R and Kwiat P 2007 *Sci. Am.* **296** 90
[13] Wheeler J A 1978 The 'past' and the 'delayed-choice' double-slit experiment *Mathematical Foundations of Quantum Theory* (Amsterdam: Elsevier) pp 9–48
[14] Wheeler J A 1980 Delayed-choice experiments and the Bohr-Einstein dialog *The American Philosophical Society and the Royal Society, Papers read at a meeting June* vol 5

Muhammad Hamza Waseem, Faizan-e-Ilahi and Muhammad Sabieh Anwar

Chapter 3

Quantum nature of light

The purpose of this chapter is to present a brief overview of quantum mechanics of what is sufficient to understand the quantum picture of light. Throughout our single-photon experiments, we will use the theoretical tools discussed in this chapter. While popular textbooks [1–3] on quantum mechanics make an excellent introduction to the subject, we follow an approach similar to the texts [4–6] because we feel they are well-suited to study photons. Mark Beck's textbook [5] covers most of the experiments discussed in this book. In fact, Beck's work has pioneered the use of photons to demonstrate quantum mechanical phenomena to avid learners of the subject and has no doubt inspired us too.

Note that this chapter is not intended as a comprehensive introduction to quantum physics. It only serves as a prelude to understanding the experiments that are the purpose of this book.

3.1 Quantum mechanical states

In the language of quantum mechanics, a state provides a complete description of a physical system. We describe the state of a physical system through a state vector, generally denoted as a ket $|\psi\rangle$. This vector resides in a complex vector space endowed with an inner product. This space is called a Hilbert space. The superposition of say two quantum states $|\psi_1\rangle$ and $|\psi_2\rangle$ is also a valid state of the same Hilbert space, and is called a superposition state. It is expressed as

$$|\psi\rangle = \alpha|\psi_1\rangle + \beta|\psi_2\rangle, \tag{3.1}$$

where α and β are complex numbers and $|\alpha|^2 + |\beta|^2 = 1$. This equality stems from the need for conserving probability. Its conjugate is described by the dual vector, denoted as the bra $\langle\psi|$, and given by

$$\langle\psi| = \alpha^* \langle\psi_1| + \beta^* \langle\psi_2|, \tag{3.2}$$

doi:10.1088/978-0-7503-3063-3ch3

where α^* and β^* represent complex conjugates of α and β, respectively. The bras $\langle\psi_1|$ and $\langle\psi_2|$ live in the dual of the original Hilbert space.

The inner product is used to determine the overlap between two vectors $|\psi\rangle$ and $|\phi\rangle$ and is denoted as

$$\langle\psi|\phi\rangle. \tag{3.3}$$

The inner product of a state with itself is real and positive. A state is said to be normalized if its inner product with itself is one. Two states are said to be orthogonal if their inner product is zero. Two normalized states which are orthogonal to each other are called orthonormal. If the component states $|\psi_1\rangle$ and $|\psi_2\rangle$ in equation (3.1) are orthonormal, the numbers α and β can be computed using the inner product as

$$\langle\psi_1|\psi\rangle = \alpha, \tag{3.4}$$

$$\langle\psi_2|\psi\rangle = \beta. \tag{3.5}$$

For a normalized $|\psi\rangle$ in equation (3.1), we have $\langle\psi|\psi\rangle = |\alpha|^2 + |\beta|^2 = 1$ where $|\alpha|^2$ and $|\beta|^2$ represent the probabilities that the state $|\psi\rangle$ will be measured to be in the states $|\psi_1\rangle$ and $|\psi_2\rangle$, respectively. The complex numbers α and β are hence termed probability amplitudes. The above treatment can be easily extended to n states inside the superposition.

All states that can be represented as state vectors are called pure states whereas states that cannot be represented as state vectors are called mixed states. They are represented by density matrices and are covered later in section 3.6.

3.2 Qubits

A qubit forms the simplest quantum system, having just two basis states. Just as a bit is the fundamental unit of information in computing, a qubit (short for **qu**antum **bit**) is the fundamental unit in quantum computing. However, unlike a classical bit, a single qubit can exist in a superposition of 0 and 1 states, usually represented as $|0\rangle$ and $|1\rangle$. We have already introduced superposition states in chapter 1. Hence, we can also write the quantum state corresponding to equation (3.1) as the qubit

$$|\psi\rangle = \alpha|0\rangle + \beta|1\rangle. \tag{3.6}$$

A qubit is actually a mathematical description of a quantum two-state system. It can be physically realized through any system having two orthogonal states, such as spin-1/2 particles [7, 8] and 2-level atoms [9, 10]. One popular approach is to employ the polarization state of a photon [11–21].

In chapter 2, the polarization state was described by a two-component vector (see table 2.1)

$$\begin{pmatrix} A \\ B\ e^{i\phi} \end{pmatrix} \tag{3.7}$$

also called the Jones vector. We can carry over the same description to a Hilbert space and identify the first and second terms in the vector as the probability amplitudes attached to the horizontal and vertical polarization states. In figure 2.3 these states are points on the diametrically opposite ends of the S_3-axis. Let's denote these orthogonal states as $|H\rangle$ and $|V\rangle$, also identified sometimes as $|0\rangle$ and $|1\rangle$. Thus the state in equation (3.6) can then be represented as

$$|\psi\rangle = A|H\rangle + B\,e^{i\phi}|V\rangle = A\begin{pmatrix}1\\0\end{pmatrix} + B\,e^{i\phi}\begin{pmatrix}0\\1\end{pmatrix} = \begin{pmatrix}A\\B\,e^{i\phi}\end{pmatrix}. \tag{3.8}$$

where A, B and ϕ are real numbers. The dual vector (i.e. bra) will then be represented by the row vector

$$\langle\psi| = (A \quad B\,e^{-i\phi}). \tag{3.9}$$

Equation (3.8) seems similar to the Jones vector representation of the polarization of classical light, covered in chapter 2. Indeed, the polarization-encoded quantum state representation for single photons is analogous to the Jones vector representation of polarization for classical sources of light. Hence, in table 3.1, we write the qubit states that are identical to the Jones vectors listed in table 2.1. However, there is one crucial subtlety. In case of classical light, we could say that A is the amplitude ('amount') of horizontally polarized light while B is the amplitude ('amount') of vertically polarized light. However, while dealing with single photons, we need to remember that A and B are not the fractions of horizontal and vertical components

Table 3.1. Polarization-encoded quantum states and the corresponding state vectors. The angles α, 45° and −45° are with respect to the horizontal axis of the lab frame of reference. In the last row, A, B and ϕ are real numbers.

Polarization	Quantum state	Column vector
Linear at angle α	$\|\alpha\rangle$	$\begin{pmatrix}\cos\alpha\\\sin\alpha\end{pmatrix}$
Horizontal (linear along x-axis)	$\|H\rangle$	$\begin{pmatrix}1\\0\end{pmatrix}$
Vertical (linear along y-axis)	$\|V\rangle$	$\begin{pmatrix}0\\1\end{pmatrix}$
Diagonal (linear at 45°)	$\|D\rangle$	$\frac{1}{\sqrt{2}}\begin{pmatrix}1\\1\end{pmatrix}$
Anti-diagonal (linear at −45°)	$\|A\rangle$	$\frac{1}{\sqrt{2}}\begin{pmatrix}1\\-1\end{pmatrix}$
Left circular	$\|L\rangle$	$\frac{1}{\sqrt{2}}\begin{pmatrix}1\\i\end{pmatrix}$
Right circular	$\|R\rangle$	$\frac{1}{\sqrt{2}}\begin{pmatrix}1\\-i\end{pmatrix}$
Arbitrary	$\|\psi\rangle$	$\begin{pmatrix}A\\B\,e^{i\phi}\end{pmatrix}$

of light. Rather, their squared moduli $|A|^2$ and $|B|^2$ determine the probability of finding the horizontal and vertical states respectively as a result of an appropriate measurement. From equations (3.4) and (3.5), these probabilities are respectively $|\langle H|\psi\rangle|^2 = |A|^2$ and $|\langle V|\psi\rangle|^2 = |B|^2$. The most general way of writing the qubit is in fact

$$|\psi\rangle = \begin{pmatrix} a \\ b \end{pmatrix} \tag{3.10}$$

where a and b are both complex numbers with $|a|^2 + |b|^2 = 1$. In this case, the dual vector becomes

$$\langle\psi| = (a^* \quad b^*). \tag{3.11}$$

In the simplest version of quantum computing, we usually perform two types of operations, unitary operations and measurement. A unitary operator usually encodes the operation of wave plates on photons whereas a measurement is a projective operation. We talk more about these operations in the following sections.

3.3 Transforming quantum states

We discussed some optical elements during our discussion of classical optics in the previous chapter. The functional description and matrix representation of those elements is no different for quantum systems. However, the mathematical notation differs a little, in that the matrices are called operators and are identified by placing 'hats' on capital letters.

Polarization-encoded quantum states are transformed through two common operations which are physically realized through single-input, single-output polarization manipulation elements. For photons, a linear polarizer carries out a linear projection operation which, when coupled with a photodetector, is usually termed 'measurement'.

A wave plate performs a unitary operation which is a transformation that preserves the norm of a state. A unitary operator can be denoted as $\hat{U}$ and is formally defined as

$$\hat{U}\hat{U}^\dagger = \hat{U}^\dagger\hat{U} = \hat{\mathbf{1}}, \tag{3.12}$$

where $\hat{U}^\dagger$ represents the adjoint of the operator and $\hat{\mathbf{1}}$ represents the identity operator which is just the 2×2 identity matrix. The unitary operation on a state can be represented as

$$\hat{U}|\psi_1\rangle = |\psi_2\rangle \tag{3.13}$$

and by definition

$$\langle\psi_2|\psi_2\rangle = \langle\psi_1|\hat{U}^\dagger\hat{U}|\psi_1\rangle = \langle\psi_1|\hat{\mathbf{1}}|\psi_1\rangle = \langle\psi_1|\psi_1\rangle, \tag{3.14}$$

i.e. the norm is preserved by an operator satisfying equation (3.12).

Table 3.2. Quantum operators for commonly used polarization manipulating elements. The axis orientations, represented by θ, are with respect to the horizontal axis of the lab frame.

Optical element	Quantum operator	Operator representation in the $\{\lvert H\rangle, \lvert V\rangle\}$ basis
Linear polarizer (axis at θ)	$\hat{O}_P(\theta)$	$\begin{pmatrix} \cos^2\theta & \sin\theta\cos\theta \\ \sin\theta\cos\theta & \sin^2\theta \end{pmatrix}$
Half-wave plate (fast axis at θ)	$\hat{O}_{HWP}(\theta)$	$\begin{pmatrix} \cos 2\theta & \sin 2\theta \\ \sin 2\theta & -\cos 2\theta \end{pmatrix}$
Quarter-wave plate (fast axis at θ)	$\hat{O}_{QWP}(\theta)$	$\begin{pmatrix} \cos^2\theta + i\sin^2\theta & (1-i)\sin\theta\cos\theta \\ (1-i)\sin\theta\cos\theta & \sin^2\theta + i\cos^2\theta \end{pmatrix}$

For a series of operations represented as $\hat{O}_1, \hat{O}_2, \cdots, \hat{O}_n$, the overall operation $\hat{O}_{\text{eff}}$ can be calculated as

$$\hat{O}_{\text{eff}} = \hat{O}_n \ldots \hat{O}_2\hat{O}_1. \tag{3.15}$$

Notice the similarity with equation (2.17).

For qubits expressed in the canonical ($\{\lvert H\rangle, \lvert V\rangle\}$) basis, the operations corresponding to wave plates and polarizers are also represented by 2×2 matrices identical to the corresponding Jones matrices. As a convenient reference for the ensuing quantum experiments, we enlist the three most commonly used operators in table 3.2.

We have briefly hinted at measurement. Let's now explore it in further detail.

3.4 Measuring quantum states

In quantum mechanics, any quantity that can be physically measured is represented by an observable (say O) and we can write a Hermitian operator $\hat{O}$ corresponding to the observable. A Hermitian operator has real eigenvalues, and its eigenvectors form an orthonormal basis. This means that we can use the eigenvectors of a Hermitian operator to represent any vector in the Hilbert space.

While measuring O in a system having the state given by $\lvert\psi\rangle$, the probability of obtaining the eigenvalue λ as a result of the measurement is given by $P(\lambda \| \psi\rangle) = \lvert\langle\lambda\lvert\psi\rangle\rvert^2$ where $\lvert\lambda\rangle$ is the corresponding eigenstate. This is called Born's rule [5]. If the outcome does appear to be λ, the measurement has created the state $\lvert\lambda\rangle$. This is the essence of a typical strong measurement.

Before making the measurement, the quantum system existed in the state $\lvert\psi\rangle$. However, once the measurement is made, the system is left projected in the state $\lvert\lambda\rangle$. The act of measurement is therefore to change the state of the system. This phenomenon is called von Neumann projection [5]. If a subsequent measurement is made in the same basis, the system is always found to be in the same state, i.e. with 100% probability.

In most of our experiments, the observable is the polarization state. If we are measuring in the canonical basis, our eigenvalues are H and V (we can assign any numbers or values to them depending on our apparatus or convention).

If we take a beam of photons generated in the state described by equation (3.10) and place a horizontally oriented linear polarizer in front of the beam, some fraction of the incoming photons will transmit through. However, *a priori*, we cannot tell which photon is transmitted and which photon is blocked. The transmission is random. We can at best talk about the probability of the transmission of photons, calculated as

$$P(H||\psi\rangle) = |\langle H|\psi\rangle|^2 = |a|^2. \tag{3.16}$$

Similarly, if the polarizer is rotated such that its transmission axis is vertically oriented, the probability of transmission will be

$$P(V||\psi\rangle) = |\langle V|\psi\rangle|^2 = |b|^2. \tag{3.17}$$

Now, if another horizontally oriented polarizer is placed after the first polarizer, we get no light transmitted because after the first projection, the system is left in the state $|V\rangle$ and

$$P(H||V\rangle) = |\langle H|V\rangle|^2 = 0. \tag{3.18}$$

In contrast, if the second polarizer is vertically polarized, we get

$$P(V||V\rangle) = |\langle V|V\rangle|^2 = 1, \tag{3.19}$$

meaning that all light coming out of the first polarizer will also be transmitted through the second polarizer. If we do the aforementioned measurements with a classical light source, we get similar results. However, instead of probabilities, we obtain proportional intensities.

So far, in the quantum description of light, we have replaced Jones vectors with state vectors, Jones matrices with operators and intensities with probabilities. While polarization optics for a single beam of light does not look very different for quantum and classical world views, two or more particles can sometimes show behavior which has no classical counterpart. For instance, two particles can become entangled, which is a purely quantum mechanical phenomenon. We now discuss such composite systems.

3.5 Composite systems and entangled states

In this section, we extend our discussion from single particles to two particles and view this bipartite system in terms of photon polarization states. Generally, if we have two independent quantum systems, we can express their composite state as the direct product given by

$$|\psi\rangle = |\psi_1\rangle \otimes |\psi_2\rangle \equiv |\psi_1\rangle|\psi_2\rangle \equiv |\psi_1\psi_2\rangle. \tag{3.20}$$

These are merely three different ways of expressing the same quantum state. The direct product takes two vectors in different Hilbert spaces and combines them to form a larger vector which expresses the two-particle state in a higher dimensional Hilbert space. In the case of polarization-encoded photons, the composite system is usually two beams of photons, and the state may be $|HH\rangle$, $|VV\rangle$, $|HV\rangle$, $|VH\rangle$, or superpositions thereof. Some examples of superposition direct product states are

$$|V\rangle \otimes \left(\frac{1}{\sqrt{2}}|H\rangle + \frac{1}{\sqrt{2}}|V\rangle\right) = \frac{1}{\sqrt{2}}|VH\rangle + \frac{1}{\sqrt{2}}|VV\rangle, \tag{3.21}$$

$$|H\rangle \otimes \left(\frac{1}{\sqrt{3}}|H\rangle + i\sqrt{\frac{2}{3}}|V\rangle\right) = \frac{1}{\sqrt{3}}|HH\rangle + i\sqrt{\frac{2}{3}}|HV\rangle, \tag{3.22}$$

$$\begin{aligned}\left(\frac{1}{\sqrt{2}}|H\rangle + \frac{1}{\sqrt{2}}|V\rangle\right) \otimes \left(\frac{1}{\sqrt{2}}|H\rangle + \frac{1}{\sqrt{2}}|V\rangle\right) &= \frac{1}{2}|HH\rangle + \frac{1}{2}|HV\rangle \\ &+ \frac{1}{2}|VH\rangle + \frac{1}{2}|VV\rangle.\end{aligned} \tag{3.23}$$

The enumerated states do not describe all the possible two-photon states. According to the superposition principle, a general composite state can be expressed as

$$|\psi\rangle = \sum_n a_n|\psi_{n1}\psi_{n2}\rangle, \tag{3.24}$$

where complex numbers a_n represent probability amplitudes and $\sum_n|a_n|^2 = 1$.

Starting with the general form in equation (3.24), we can make our way backwards and attempt to factorize the state into a direct product of component states. This may not always be possible. States that are not expressible as direct products of component states are called entangled states [5, 22]. Mathematically, these states may not seem to have profound implications. However, physically, these states represent some of the most bizarre characteristics of quantum mechanics. We will experimentally play with such systems in the forthcoming chapters. For example, we will generate one of the four famous Bell states [22], which are given by

$$|\phi^+\rangle = \frac{1}{\sqrt{2}}(|HH\rangle + |VV\rangle), \tag{3.25}$$

$$|\phi^-\rangle = \frac{1}{\sqrt{2}}(|HH\rangle - |VV\rangle), \tag{3.26}$$

$$|\psi^+\rangle = \frac{1}{\sqrt{2}}(|HV\rangle + |VH\rangle), \tag{3.27}$$

$$|\psi^-\rangle = \frac{1}{\sqrt{2}}(|HV\rangle - |VH\rangle). \tag{3.28}$$

All of these are entangled states. Neither of these states can be expressed as a direct product of two independent qubits. It is not possible to factorize them.

Furthermore, in each Bell state, the two-photon system can be viewed as simultaneously *existing* in two states. For example, $|\phi^+\rangle$ is in $|HH\rangle$ and $|VV\rangle$ simultaneously. If we measure one of the two photons of $|\phi^+\rangle$ and find it in the state $|H\rangle$, the other photon is sure to be found in the state $|H\rangle$ whether or not we choose to measure it. Similarly, if one of the photons is measured in $|V\rangle$, the other will project or 'collapse' to $|V\rangle$. In other words, the state measurement of one party instantaneously determines the state of the other party. This is the essence of entanglement.

3.6 Mixed states and the density matrix

All the states we have described so far can be represented as state vectors $|\psi\rangle$ and are called pure states. Even the entangled states expressed in equations (3.25)–(3.28) are pure states. No doubt they are superpositions but they are still quantum mechanically pure. In contrast, sometimes, the system may be randomly prepared as a mixture of two possible states (say $|HH\rangle$ and $|VV\rangle$). In this case, the system is in an incoherent *statistical mixture* of the two possible states. For instance, the state $|HH\rangle$ may be added to the mixture with a probability of 50% and the state $|VV\rangle$ may be introduced into the mixture with a probability of 50%. Such states are called mixed states and cannot be represented as kets. This mixture is unequivocally different from the pure superposition state $(|HH\rangle + |VV\rangle)/\sqrt{2}$.

Mixed states can be represented using the density matrix formalism [22], in which the density operator $\hat{\rho}$ represents a generalized quantum state. This operator can be used to describe not just pure states but also mixed states, unlike the state vector which describes only pure states. If a mixture of states $|\psi_j\rangle$ is prepared with the probabilities $P(|\psi_j\rangle) = p_j$ respectively, the density operator can be expressed as

$$\hat{\rho} = \sum_j p_j |\psi_j\rangle\langle\psi_j|. \tag{3.29}$$

The probabilities must be real positive numbers in the range $0 \leqslant p_j \leqslant 1$ and must be normalized $\sum_j p_j = 1$. The states that are part of mixed states need not be orthogonal or form a basis. The component states are any states that the system can be prepared in. For a pure state $|\psi\rangle$, the density operator of equation (3.29) simply reduces to

$$\hat{\rho} = |\psi\rangle\langle\psi|. \tag{3.30}$$

This implies that the density operator of a mixed state is a probability-weighted superposition of density operators of pure states. The normalization condition for a density matrix then requires that $\mathrm{Tr}(\hat{\rho}) = 1$. Just like any other operator for a system with discrete basis, the density operator can be expressed as a matrix.

Let us look at few examples of density matrices. The density matrix for the pure Bell state given in equation (3.25) can be computed as

$$\begin{aligned}\hat{\rho}_{\phi^+} &= |\phi^+\rangle\langle\phi^+| \\ &= \frac{1}{\sqrt{2}}(|HH\rangle + |VV\rangle)\frac{1}{\sqrt{2}}(\langle HH| + \langle VV|) \\ &= \frac{1}{2}|HH\rangle\langle HH| + \frac{1}{2}|HH\rangle\langle VV| + \frac{1}{2}|VV\rangle\langle HH| \\ &+ \frac{1}{2}|VV\rangle\langle VV| \\ &= \frac{1}{2}\begin{pmatrix} 1 & 0 & 0 & 1 \\ 0 & 0 & 0 & 0 \\ 0 & 0 & 0 & 0 \\ 1 & 0 & 0 & 1 \end{pmatrix}.\end{aligned} \tag{3.31}$$

The matrix form in the last line can be written with the help of some basic matrix algebra. For example, using the vector forms of $|H\rangle$ and $|V\rangle$ in table 3.1, we can write the vector $|HH\rangle$ as

$$|HH\rangle = \begin{pmatrix} 1 \\ 0 \end{pmatrix} \otimes \begin{pmatrix} 1 \\ 0 \end{pmatrix} = \begin{pmatrix} 1 \\ 0 \\ 0 \\ 0 \end{pmatrix} \tag{3.32}$$

with the dual bra

$$\langle HH| = (1 \;\; 0 \;\; 0 \;\; 0). \tag{3.33}$$

Finally, juxtaposing the bra and the ket yields

$$|HH\rangle\langle HH| = \begin{pmatrix} 1 & 0 & 0 & 0 \\ 0 & 0 & 0 & 0 \\ 0 & 0 & 0 & 0 \\ 0 & 0 & 0 & 0 \end{pmatrix}. \tag{3.34}$$

This process can be repeated to finally construct the matrix in equation (3.31).

In contrast, the density matrix for a mixed two-photon state, in which half the time the photons are detected in the state $|HH\rangle$ and half the time they are detected in the state $|VV\rangle$, is given by

$$\begin{aligned}\hat{\rho}_{\text{mix}} &= \frac{1}{2}|HH\rangle\langle HH| + \frac{1}{2}|VV\rangle\langle VV| \\ &= \frac{1}{2}\begin{pmatrix} 1 & 0 & 0 & 0 \\ 0 & 0 & 0 & 0 \\ 0 & 0 & 0 & 0 \\ 0 & 0 & 0 & 1 \end{pmatrix}.\end{aligned} \tag{3.35}$$

There could be an infinite variety of mixed states. Another important two-qubit state is the Werner state [23], represented by

$$\hat{\rho}_W = \varepsilon|\psi_{\text{ent}}\rangle\langle\psi_{\text{ent}}| + (1 - \varepsilon)\frac{\hat{\mathbf{1}}}{4}, \tag{3.36}$$

where $|\psi_{\text{ent}}\rangle\langle\psi_{\text{ent}}|$ represents the density matrix of a maximally entangled state, given by one of the four Bell states described in the previous section, and $\hat{\mathbf{1}}/4$ (where $\hat{\mathbf{1}}$ is the 4×4 identity matrix) represents the density matrix of maximally mixed two-qubit state. The variable ε represents the probability the photons are in the state $|\psi_{\text{ent}}\rangle\langle\psi_{\text{ent}}|$ and $(1 - \varepsilon)$ represents the probability the photons are in the maximally mixed state $\hat{\mathbf{1}}/4$.

For a pure state given by $|\psi\rangle\langle\psi|$, it is evident that $\text{Tr}(\hat{\rho}^2) = \text{Tr}(\hat{\rho}) = 1$. In contrast, for mixed states, it can be seen that $\hat{\rho}^2 \neq \hat{\rho}$ and $\text{Tr}(\hat{\rho}^2) < 1$. So, $\text{Tr}(\hat{\rho}^2)$ can be used as a figure of merit for the 'purity' of a system. The closer it is to 1, the purer the state is [5]. For a completely mixed two-qubit state, $\text{Tr}(\hat{\rho}^2) = 1/4$. The denominator is 4 which is the dimensionality of the system.

As a last example, consider an equal mixture of the states $|HH\rangle$, $|HV\rangle$, $|VH\rangle$ and $|VV\rangle$. The density matrix of this incoherent mixture will be

$$\frac{1}{4}|HH\rangle\langle HH| + \frac{1}{4}|HV\rangle\langle HV| + \frac{1}{4}|VH\rangle\langle VH| + \frac{1}{4}|VV\rangle\langle VV| = \frac{\hat{\mathbf{1}}}{4} \tag{3.37}$$

which is maximally mixed. Each of the constituents of the mixture is a separable state of the two photons. Interestingly, we can also show that if we compose an equal mixture of non-separable, entangled Bell states, we could still obtain the maximally mixed state. Try to see if you could verify the following:

$$\frac{1}{4}|\phi^+\rangle\langle\phi^+| + \frac{1}{4}|\phi^-\rangle\langle\phi^-| + \frac{1}{4}|\psi^+\rangle\langle\psi^+| + \frac{1}{4}|\psi^-\rangle\langle\psi^-| = \frac{\hat{\mathbf{1}}}{4}. \tag{3.38}$$

3.7 Photon statistics

Apart from multi-photon entangled systems, another aspect in which single photons remarkably differ from coherent light is in the statistics of photodetection. To illustrate, photodetection statistics for coherent light with a stable intensity are described by a Poisson distribution [4], represented as

$$p(x) = \frac{n^x}{x!}e^{-n} \quad x = 0, 1, 2, \dots, \tag{3.39}$$

where x represents the number of photons successfully detected in a given time duration, n represents the mean photodetections while the standard deviation is given by $\Delta n = \sqrt{n}$. If there are thermal fluctuations in the light source, a larger standard deviation is seen and such light is said to exhibit super-Poissonian statistics ($\Delta n > \sqrt{n}$). However, a beam of single photons exhibits sub-Poissonian ($\Delta n < \sqrt{n}$)

statistics [4], which has no classical counterpart. The creation and detection of sub-Poissonian light is quite difficult but a true signature of the quantum nature of light.

In the next chapter, we will discuss experiments related to measuring the statistics and polarization of single photons. The theoretical toolbox developed in this chapter will prove to be very useful all the while.

References

[1] Griffiths D J 2016 *Introduction to Quantum Mechanics* (Cambridge: Cambridge University Press)

[2] Shankar R 2012 *Principles of Quantum Mechanics* (Berlin: Springer)

[3] Dirac P A M 1981 *The Principles of Quantum Mechanics* (Oxford: Oxford University Press)

[4] Fox M 2006 *Quantum Optics: An Introduction* (Oxford: Oxford University Press)

[5] Beck M 2012 *Quantum Mechanics: Theory and Experiment* (Oxford: Oxford University Press)

[6] Lvovsky A 2018 *Quantum Physics: An Introduction Based on Photons* (Berlin: Springer)

[7] Cory D G, Fahmy A F and Havel T F 1997 *Proc. Natl. Acad. Sci. U.S.A.* **94** 1634

[8] Weinstein Y S, Pravia M, Fortunato E, Lloyd S and Cory D G 2001 *Phys. Rev. Lett.* **86** 1889

[9] Monroe C 2002 *Nature* **416** 238

[10] Schmidt-Kaler F *et al* 2003 *Nature* **422** 408

[11] White A G, James D F, Eberhard P H and Kwiat P G 1999 *Phys. Rev. Lett.* **83** 3103

[12] Sanaka K, Kawahara K and Kuga T 2001 *Phys. Rev. Lett.* **86** 5620

[13] Mair A, Vaziri A, Zeilinger A and Weihs G 2001 *Nature* **412** 3123

[14] Nambu Y, Usami K, Tsuda Y, Matsumoto K and Nakamura K 2002 *Phys. Rev.* A **66** 033816

[15] Yamamoto T, Koashi M, Özdemir Ş K and Imoto N 2003 *Nature* **421** 343

[16] Sergienko A *et al* 2003 Entangled-photon state engineering *Proc. of the Sixth Int. Conf. on Quantum Communication, Measurement and Computing (QCMC) (Rinton Princeton)* pp 147–52

[17] Pittman T, Fitch M, Jacobs B and Franson J 2003 *Phys. Rev.* A **68** 032316

[18] O'Brien J L, Pryde G J, White A G, Ralph T C and Branning D 2003 *Nature* **426** 264

[19] Marcikic I, De Riedmatten H, Tittel W, Zbinden H and Gisin N 2003 *Nature* **421** 509

[20] Pearson B J and Jackson D P 2010 *Am. J. Phys.* **78** 471

[21] Brody J and Selton C 2018 *Am. J. Phys.* **86** 412

[22] Nielsen M A and Chuang I 2010 *Quantum Computation and Quantum Information* (Cambridge: Cambridge University Press)

[23] Werner R F 1989 *Phys. Rev.* A **40** 4277

IOP Publishing

Quantum Mechanics in the Single Photon Laboratory

Muhammad Hamza Waseem, Faizan-e-Ilahi and Muhammad Sabieh Anwar

Chapter 4

Experiments related to the quantum nature of light

Experiments based on statistics of photons and single-photon states have been explored as an effective tool to study and teach quantum mechanics in the instructional laboratory, where simplicity, affordability and modularity are important concerns. Important work in this regard has been championed by experimental physics groups headed by Beck [1–4], Galvez [1, 2, 5–7] and Lukishova [8–10] and are also surveyed in chapter 1.

The content of this short textbook is motivated by all of these studies, especially Beck's work [1–4]. A general overview of this 'single-photon lab' can be visualized through figure 4.1, which schematically depicts three major components of the lab: setups for the optics, photon counting, and post-processing of photodetection statistics. Furthermore, figure 4.2 shows a bird's-eye view of the lab actually constructed by the authors of this book in PhysLab, LUMS, Pakistan.

For the sake of convenience, we will refer to our tabletop single-photon experiments through the following nomenclature:

- Q1: Spontaneous parametric downconversion.
- Q2: Proof of existence of photons.
- Q3: Estimating the polarization state of single photons.
- Q4: Visualizing the polarization state of single photons.
- Q5: Single-photon interference and quantum eraser.

In this chapter, we will discuss the instrumentation details common to all our experiments. Moreover, we will elaborate experiments Q1–Q5, which focus on quantum properties of a single beam of photons. More precisely, experiment Q1 is about setting up a source for creating single photons using the process of downconversion. Experiment Q2 shows the granular nature of light. Often dubbed as a proof of existence of photons, this experiment serves as a confirmation of the successful generation of single-photon states. Experiments Q3 and Q4 estimate and

doi:10.1088/978-0-7503-3063-3ch4

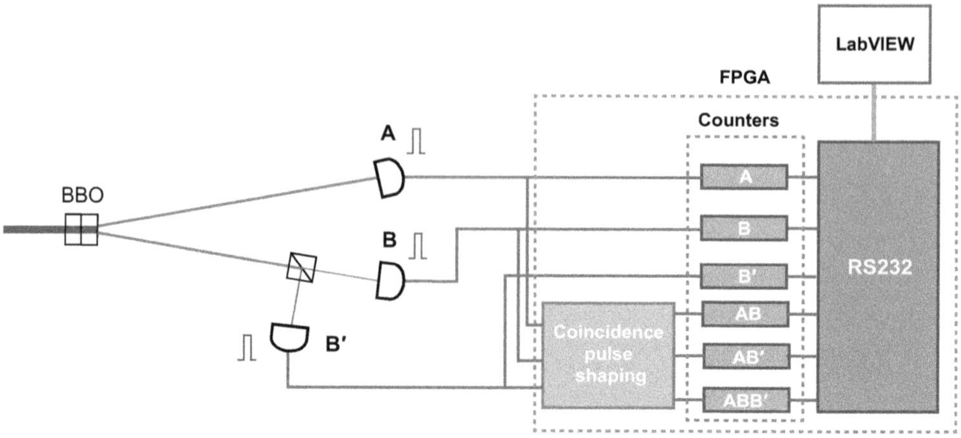

Figure 4.1. The three major aspects of the single-photon experiments include the optical setup, photon counting mechanism, and post-processing system of the photon statistics. The elements before A, B and B' form the optical setup; A, B and B' detect the photons and the FPGA and computer perform the post-processing. The sample optical setup here shows BBO crystals, a polarizing beam splitter and three detectors A, B and B'. The FPGA detects and counts single photon and coincidence photon pulses and sends the photocounts periodically via serial communication to a LabVIEW program running on a computer. The pulse next to A, B and B' detectors shows a TTL voltage (= 5 V) pulse of duration approximately 20 ns.

visualize, respectively, the polarization state of single photons. Finally, experiment Q5 brings home wave-particle duality by exploring the interference of single photons and quantum erasure.

Since the experiments are modular, we have ordered them such that the subsequent experiments build up on the setup of the preceding ones. Before we jump to the experiments, we need to discuss some technical aspects of the instrumentation which forms the backbone of all our experiments.

4.1 General components of the lab

For convenience, we divide this discussion into three groups: the light source, the light detection system and the coincidence counting unit. The individual parts we have employed in our experiments are listed in appendix A.

4.1.1 Light source

In all our experiments, photon pairs are produced through the nonlinear process of downconversion. This process is discussed in sufficient detail in experiment Q1. Downconversion is a highly inefficient process and hence the experiments need to be pumped with a sufficiently bright light source. Such a laser will also help in doing the optical alignment quickly and consequently, the actual experiments could be completed in a few minutes. We use a vertically polarized, 405 nm violet–blue laser, which is quite convenient for these experiments because of its turn-key operation and high nominal power of 50 mW, which can be controlled by rotating a knob (figure 4.3(a)). Before falling on the downconversion crystal, the laser is

Figure 4.2. The single-photon lab in PhysLab, LUMS. The optical components are set up on the optical table. Single-photon detectors transmit TTL pulses for each photodetection. The pulses are counted by an FPGA, which transmits photon count information to a PC, where counts are monitored via a LabVIEW program. The single-photon experiments are performed with lights turned off.

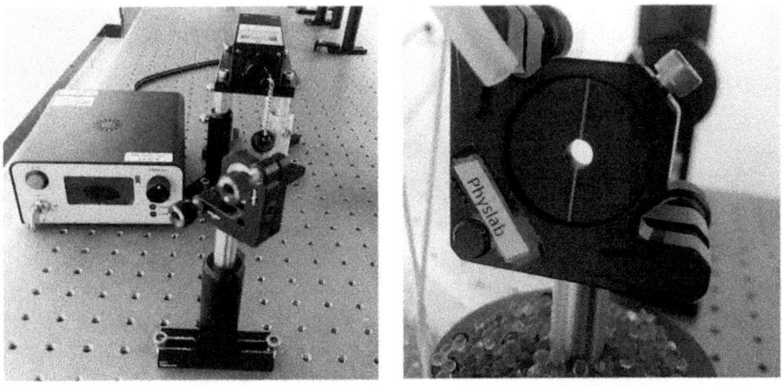

(a) (b)

Figure 4.3. (a) A 50 mW, 405 nm, turn-key laser is used as the pump laser for the single-photon experiments. (b) Mounted BBO crystals, held in a mirror mount. The white line shows the axis of the crystal. To protect the crystals from moisture, silica gel is placed below the mount and a small pipe is used to provide a gentle flow of nitrogen.

made to pass through a half-wave plate (HWP), which helps adjusting the laser polarization.

Downconversion is achieved in two stacked β-barium borate (BBO, BaB_2O_4) crystals, cut for type-I downconversion (figure 4.3(b)). Since they are hygroscopic, the crystals are protected from moisture by placing silica gel beneath them and maintaining a gentle flow of nitrogen over the crystal mount. Moreover, when the crystals are not being used, they are stored in a desiccant jar.

The performance or quality of the light source can be gauged in terms of single and coincidence count rates, which can be measured in counts per second (cps). In our various experiments, we achieved maximum single-photon detection rates of around 70 000 cps and coincidence count rates of 3000 cps between the signal and idler beams. We will shortly describe these beams.

4.1.2 Light detection

The light collection optics that we use helps alignment as well as rejection of background light. Optical fibres help make the system flexible and efficient. The collection optics and detection schemes described in this section will be used identically for all the detectors in all the experiments.

As shown in figure 4.4(a), the downconverted photons are collected with fibre-coupling lenses, coupled into multimode fibre optic cables having fibre-coupling (FC) connectors at both ends and directed into a single-photon counting module (SPCM) where they are detected. In front of the collection lenses are long pass optical filters. These filters block light having wavelengths smaller than 780 nm and transmit light of higher wavelengths. These filters are required to block the ambient or scattered light which can not only distort the photon statistics but can also damage the SPCM.

The lens is packaged as pre-aligned so that there is no need to align the fibre with the lens. We connect the other end of the fibre with a fibre-to-fibre coupler, which helps coupling the light into another fibre (figure 4.4(b)). This arrangement is very useful and gives us the flexibility to swap the connections between the different

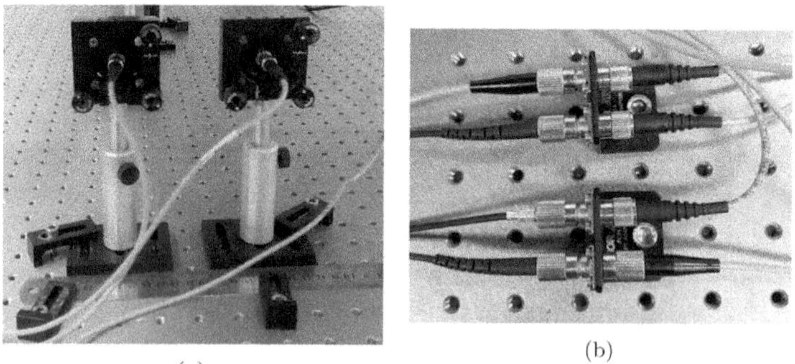

(a) (b)

Figure 4.4. (a) Fibre-coupling lenses are held in kinematic mounts to collect the downconverted photons. The photons are coupled into multimode fibre optic cables, the other end of which are connected to (b) fibre-to-fibre couplers, whose role is to couple light from one fibre into another.

coupling lenses and the different detectors. This is useful for setting up and testing the coincidence counting system. This scheme is also helpful for connecting a fibre-coupled alignment laser in order to propagate it backwards through the coupling lens for alignment purposes (more on this in the alignment sections).

Kinematic mounts are used to hold both the fibre coupled collimation lenses and the long pass filters. The mounts are flexible and efficient for alignment purposes. They have knobs which can adjust the horizontal and vertical tilt of the coupling lenses.

The second fibre takes the photons to the SPCM. We use an SPCM which has four independent channels of avalanche photodiodes (APDs) (figure 4.5), which can detect photons between wavelengths 400 nm and 1060nm and have a dead time of 50 ns between pulses. The APD modules are optimized for a peak photon detection efficiency of over 60% at wavelength 650 nm.

For each photodetection, an output TTL pulse (5 V) of about 20–25 ns is produced at the BNC output port of the corresponding channel. A high speed oscilloscope was used to view this pulse (figure 4.6). The APDs set are powered by three power supplies of 2 V, 5 V and 30 V.

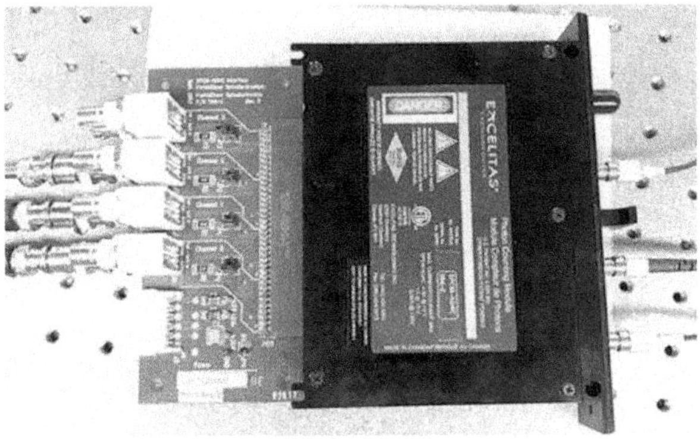

Figure 4.5. The single-photon counting module (SPCM), which has four independent detectors, three of which are in use in this picture. The photodetection output pulses are transmitted via the BNC terminals.

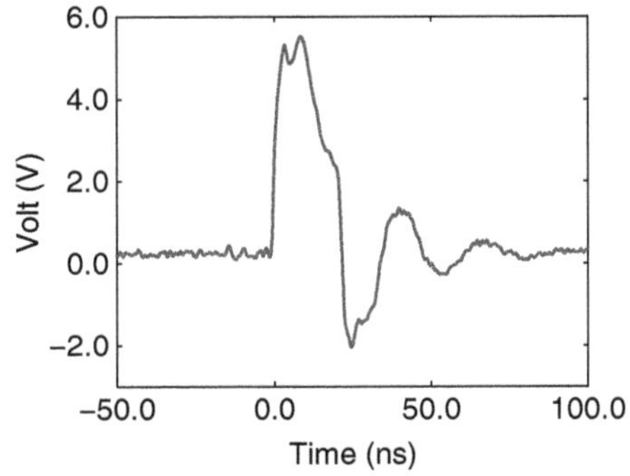

Figure 4.6. Signature pulse of a single photodetection by an APD.

It is important to keep the APDs safe from receiving ambient light; otherwise they can be easily damaged due to excess of photons. For this purpose, the experiments need to be conducted with the room lights turned off. Thermal fluctuations can cause the APDs to register photons even when there is no light incident on the APDs. These are called dark counts. For the four detectors, the manufacturer mentioned the dark counts to be (311 ± 18) cps, (365 ± 19) cps, (243 ± 16) cps and (344 ± 19) cps, while our measurements at (333 ± 18) cps, (352 ± 19) cps, (242 ± 16) cps and (339 ± 18) cps respectively were in the prescribed range.

4.1.3 Coincidence counting unit

Coincidence counting means detecting two or more particles simultaneously at different detectors [11]. Widely employed in experimental physics, this technique plays an essential role in experiments pertaining to quantum optics. Coincidence detection and photon counting lie at the heart of investigating and making use of the quantum characteristics of correlated light sources. As we will see in this book, most of these experiments demand counting of only two-fold coincidences while others require counting of multiphoton coincidences for more than two detectors [12–14].

Traditionally, coincidence counting has been mostly performed using time-to-amplitude converters (TACs), where each TAC allows to count one coincident photon pair [15]. Hence, it is obvious that coincidence counting becomes very difficult and costly for multiphoton experiments. Moreover, the maximum rate of TAC-based coincidence counting is severely capped by the conversion time involved in each start or stop event. In the last two decades, these problems have been addressed and many solutions have been proposed focusing on particular applications, including quantum information processing [16], fluorescence measurements [17, 18], x-ray microscopy [19], and physics education [11, 20–22]. For performing experiments in quantum mechanics or quantum optics, a cost-effective solution for coincidence counting is based on the field-programmable gate array (FPGA) [23, 24]. We develop a minimalist version of such a coincidence counting unit (CCU), as discussed below. We break our discussion into hardware and software parts.

The hardware

The pulses generated by the SPCM are counted through a CCU based on Nexys 2 FPGA[1]. This device is shown in figure 4.7. An FPGA is an integrated circuit with a large number of logic blocks that can be programmed to perform the desired operations. We use an FPGA that has an internal clock of 50 MHz, which enables an effective coincidence window of 2 × 20 ns and hence provides an efficient, cost-effective solution for simultaneous counting of single and coincident photons from a number of photodetectors.

The CCU counts the single and the coincident photodetection events simply by detecting the rising edges in the pulse-based signals. The photons are considered

[1] For background information on FPGAs, see appendix B.

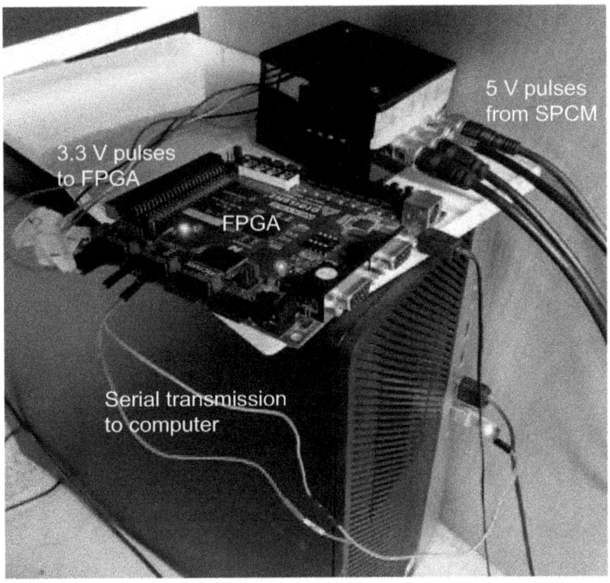

Figure 4.7. The coincidence counting unit consists of an FPGA, which can simultaneously count up to four single photon detections and up to four coincidence photon detections. The black box consists of potential dividers which step down the 5 V signals from the SPCM to 3.3 V signals. The FPGA transmits the photodetection counts to the computer through serial communication via a USB–TTL converter.

coincident if they arrive at different detectors within a specified coincidence window of time (i.e. 40 ns in our case). Coincidence counting calculations are also performed by hardware logic implemented on the FPGA, which communicates all coincidence and single count information to a computer. Finally, using a LabVIEW-based interface on the computer, the experimenter can monitor the single detection counts and coincidence counts in the form of numbers and plots.

The logic inputs of the FPGA are rated at 3.3 V. Therefore, 5 V pulses produced by the SPCM are transmitted to the FPGA through a potential divider box. The FPGA is programmed to detect two-detector and three-detector coincidences from different detectors. To perform this coincidence detection, the pulses for the coincidence in question are simply ANDed together, producing an output only if the pulses overlap in time (figure 4.8). For two pulses, this overlap only occurs if one pulse arrives after another pulse within the width of the first pulse. The coincidence window is directly proportional to width of the pulses. In fact, due to the independence of which pulse arrives first, the true coincidence window is exactly twice the width of the pulses. In our case, this window is 40 ns.

The FPGA is programmed to use eight 32-bit counters, four of which count single photons from four detectors while the rest count coincidences. Not all counters are used in each experiment. Moreover, the type of coincidence detection may vary from experiment to experiment. Therefore, these slight changes are made in the FPGA program before performing a different experiment. The fully annotated code of the FPGA is listed in appendix C.

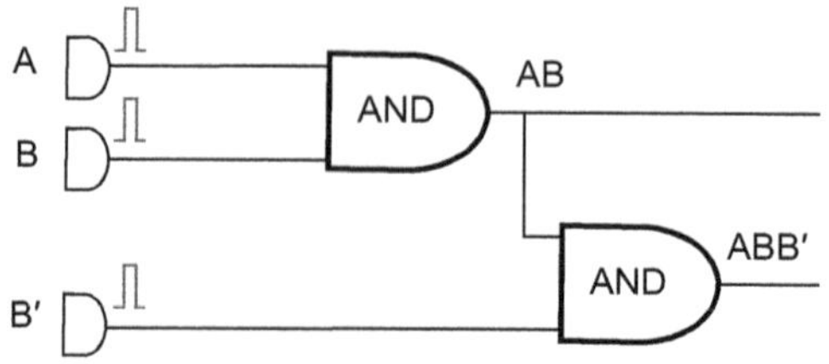

Figure 4.8. Schematic description of coincidence logic programmed on the FPGA. To determine the coincidence signal of two or more detectors, the single detector signals are simply ANDed and the resulting signal is transmitted to a regular counter in the FPGA.

The coincidence counts and single counts information are transferred to the computer through RS-232 serial communication and all the counters are reset every 0.1 s. For monitoring these photocounts, we use a LabVIEW program based on the one developed in Beck's group[2].

The software

The LabVIEW program receives the photocount data in the form of a bit-stream, decodes it into a readable format and displays the single photon and coincidence counts in the form of numbers and plots. For each experiment, we make a different variation of the program depending on the required information.

The front panel of the program (figure 4.9(a)) is minimalistic and the common functions in all the versions of the program are a STOP button, and two input parameters. The Counter Port input specifies the hardware component in communication with the LabVIEW software, which in this case is the RS-232 serial connection. The Update Period input is how often the displayed data is refreshed on the interface, and is selected as increments of 0.1 s. This is because 0.1 s is the smallest update period due to the read-out rate of the FPGA. Figure 4.9(b) shows the block diagram of the LabVIEW program.

The overall inventory as well as optical components for individual experiments are listed in appendix A. Finally, we now discuss the first quantum optics experiment.

4.2 Q1: Spontaneous parametric downconversion

At the heart of all our quantum optics experiments is a process called spontaneous parametric downconversion (SPDC) [20, 25–27]. The process is called downconversion since the frequency of the output beams is smaller than the frequency of the input beam. This is a nonlinear process since it changes the frequency of the light beam, in contrast to the linear optical processes studied in chapter 2, which could change many properties of light such as the intensity, phase or polarization but not the frequency.

In SPDC, schematically illustrated in figure 4.10, a pump (p) photon of one frequency produces two photons (signal (s) and idler (i)) of half the original

[2] http://people.reed.edu/beckm/QM/labview/labview.html.

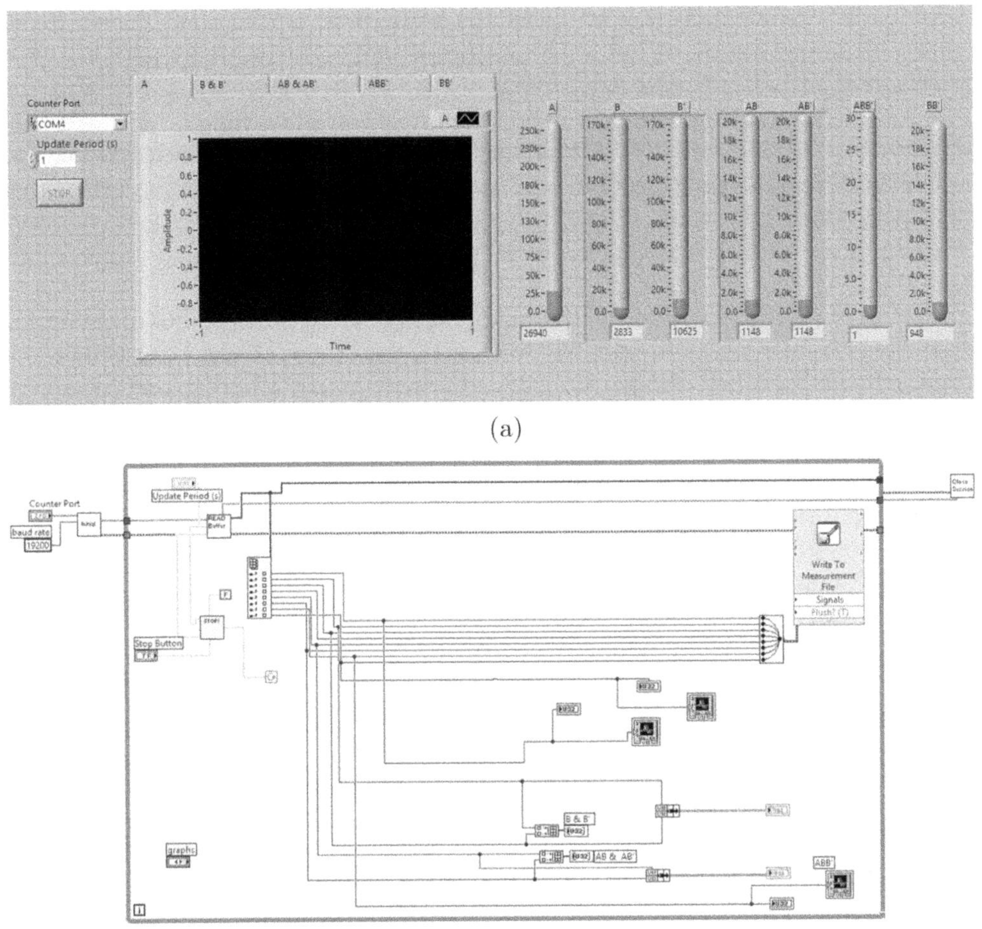

Figure 4.9. (a) A screenshot of the front panel of the LabVIEW program for Experiment Q2. Counter Port is used to select the hardware port for serial communication. Update Period helps select the data acquisition rate in increments of 0.1 s. (b) A screenshot of the block diagram of the same program.

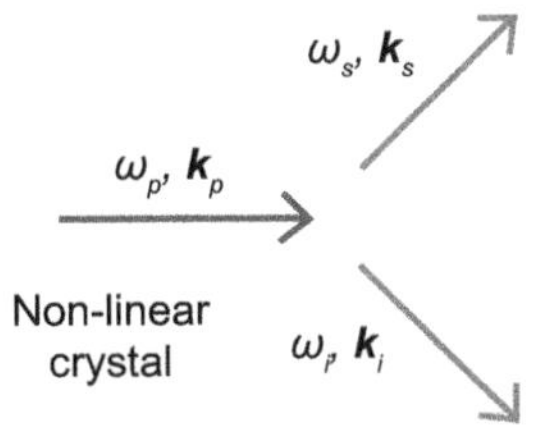

Figure 4.10. Concept diagram of type-I spontaneous parametric downconversion. A photon of frequency ω_p and wave vector $\mathbf{k}_p$ is absorbed and two photons of frequencies ω_s and ω_i and wave vectors $\mathbf{k}_s$ and $\mathbf{k}_i$ are released.

frequency. The downconverted photon pairs are used as a light source to study the peculiar quantum nature of light. The process is dubbed 'spontaneous' since there are no signal and idler beams in the input but are created spontaneously. In fact, in SPDC, an electron is excited by the incident pump photon. As the electron returns to the ground state, it emits two photons rather than one. Furthermore, SPDC is a parametric process since it depends on the electric fields (polarizations) rather than intensities of the light beams. Therefore, a definite phase relationship exists between the pump beam and the downconverted beams. The phase matching requirements are outlined in the next section.

SPDC has a number of merits that make it effective in tabletop quantum optical experiments. While SPDC is extremely inefficient, it is more efficient as compared to atomic cascade used by Grangier *et al* [28] and much easier to implement. Moreover, the downconverted photons are always emitted in pairs. Thus detection of one photon tags the presence of its sibling. This allows SPDC to be used as a heralded single-photon source. Owing to its relative simplicity, SPDC has been used in correlated-photon pair experiments such as tests of existence of photons [25], single-photon interference [3], quantum erasure [29, 30], state measurement of single photons [3], quantum state tomography [31], and tests of Bell inequalities [32–34].

4.2.1 The downconversion crystal and phase-matching

For SPDC, we used two stacked BBO crystals, each of dimensions $5 \times 5 \times 0.5$ mm. They are mounted in such a way that the axis of one is rotated 90° relative to the other. This allows for downconversion of both horizontally and vertically polarized photons. The BBO crystals are cut for type-I SPDC to produce photon pairs with linear polarization parallel to each other but orthogonal to the polarization of the input beam. This means that the downconversion relations for the two crystals can be described separately, as

$$|V\rangle \longrightarrow |HH\rangle, \tag{4.1}$$

$$|H\rangle \longrightarrow |VV\rangle. \tag{4.2}$$

Moreover, if the pump polarization is oriented at an angle θ with respect to the vertical, the vertical and horizontal components of the input beam will respectively result in $\cos^2\theta|HH\rangle$ and $\sin^2\theta|VV\rangle$. At this stage, we are only looking at the two downconversion crystals separately. However, when we are interested in the polarization correlations of the two beams, an important subtlety arises about this assembly of two BBO crystals [26]. If one cannot distinguish between the photon pairs produced by the two crystals, the two-photon state is entangled. We will discuss this in detail in chapter 5 where we talk about entanglement and nonlocality.

The direction taken by the output photons is determined by the angle formed by the optic axis (OA) of the crystal with the direction of propagation of the pump beam. This angle is called the phase matching angle θ_m.

In our quantum optical experiments, for the pump beam we use a laser having a wavelength of approximately 405 nm. The downconverted beams are of wavelength

810 nm, which is twice the wavelength and hence half the frequency of the pump beam. To ensure the separation of the two downconverted beams, they are made to exit the BBO crystal each making an angle of 3° with respect to the pump beam. Therefore, the signal beam comes out of the downconversion crystal making an angle of 3° with respect to the pump and likewise the idler beam exists the downconversion source making an angle of 3° but on the opposite side of the pump beam. In fact, the absolute angles between the pump and the downconverted beams do not matter. Only the relative angles are important. Therefore, it can be observed that both the downconverted beams make cones which surround the pump beam.

Shown in figure 4.11, conservation of energy requires that

$$\omega_p = \omega_s + \omega_i, \tag{4.3}$$

whereas conservation of momentum requires that

$$\mathbf{k}_p = \mathbf{k}_s + \mathbf{k}_i. \tag{4.4}$$

The frequencies and wave vectors of the optical beams are not independent of each other but are in fact related. For the pump wave, for instance, we have the relation

$$k_p = \frac{n_p \omega_p}{c}, \tag{4.5}$$

where n_p denotes the refractive index of the downconversion crystal at the pump frequency. This is called a dispersion relationship. Similar dispersion relations exist for the signal and idler waves. Therefore, equations (4.4) and (4.5) lead to the component conditions

$$n_p \omega_p = n_s \omega_s \cos\theta_s + n_i \omega_i \cos\theta_i \text{ and} \tag{4.6}$$

$$0 = n_s \omega_s \sin\theta_s + n_i \omega_i \sin\theta_i. \tag{4.7}$$

The downconverted photons come out of the crystal at a range of wavelengths and angles. However, for our experiments, we consider the photons which come out in the horizontal plane and for which $\omega_s = \omega_i = \omega_p/2$, $n_s = n_i$ and $\theta_s = \theta_i$. Therefore, equation (4.6) becomes

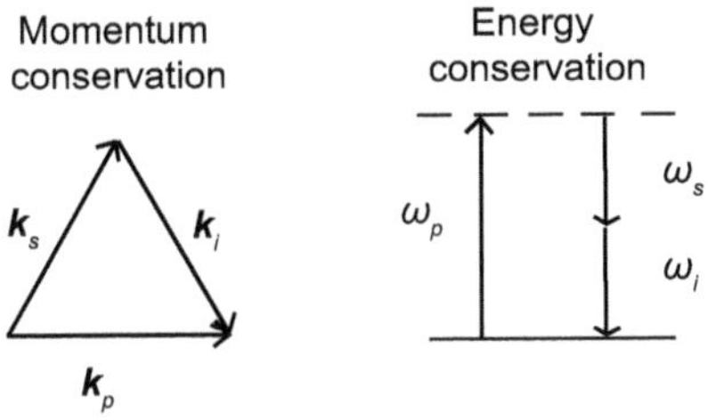

Figure 4.11. Momentum $\hbar\mathbf{k}$ and energy $\hbar\omega$ are conserved in downconversion. Subscripts p, s and i stand for pump, signal and idler, respectively.

$$n_p = n_s \cos \theta_s. \tag{4.8}$$

It is not possible to satisfy this equation in an isotropic medium because, for normal dispersion, $n_p > n_s$ when $\lambda_p < \lambda_s$, leading to unphysical values of the cosine function. However, this problem can be overcome with a BBO crystal, which is a uniaxial birefringent crystal, having *two* indices of refraction. Light is said to have ordinary polarization if it is polarized perpendicular to the optic axis of the crystal. In this case, it has an ordinary index of refraction denoted as n_o. On the other hand, for light having polarization parallel to the optic axis, there is an extraordinary index of refraction denoted as n_e. Plotted in figure 4.12, the indices of refraction of the BBO crystal are provided by the crystal grower as[3]

$$n = \left(A + \frac{B}{\lambda^2 + C} + D\lambda^2\right)^{1/2}, \tag{4.9}$$

where the wavelength λ is in μm and the constants for n_o and n_e are given by
- $A_o = 2.735\,9$
- $B_o = 0.018\,78\ \mu\text{m}^2$
- $C_o = -0.018\,22\ \mu\text{m}^2$
- $D_o = -0.013\,54\ \mu\text{m}^{-2}$
- $A_e = 2.375\,3$
- $B_e = 0.012\,24\ \mu\text{m}^2$
- $C_e = -0.016\,67\ \mu\text{m}^2$
- $D_e = -0.015\,16\ \mu\text{m}^{-2}$.

If light polarization is at an angle with respect to the optic axis, the index of refraction is modified. This effective index of refraction n_{eff} depends on the phase matching angle, denoted as θ_m, between the propagation direction and the optic axis. This relation is given by [35]

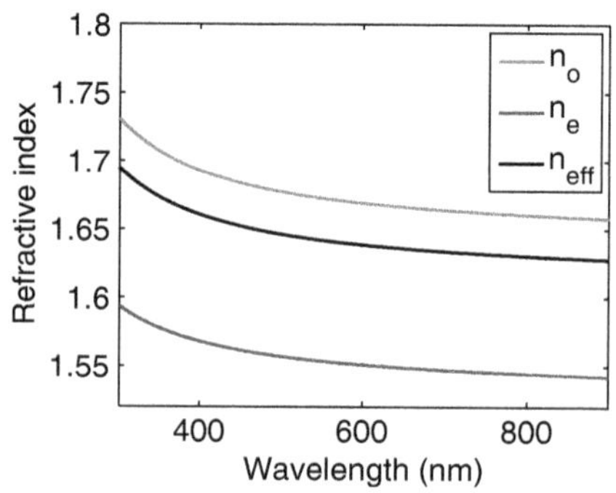

Figure 4.12. Refractive indices of BBO versus wavelength of light. The effective index n_{eff} is tuned between ordinary index n_o and extraordinary index n_e by tuning the phase matching angle.

[3] https://www.unitedcrystals.com/BBOProp.html.

$$n_{\text{eff}}(\theta_m) = \left(\frac{\cos\theta_m^2}{n_o^2} + \frac{\sin\theta_m^2}{n_e^2}\right)^{-1/2}. \tag{4.10}$$

Hence, the effective refractive index can be 'tuned' between n_e and n_o by tuning θ_m. In this way, we can get the desired index of refraction for the pump beam which is required to satisfy equation (4.8).

In type-I phase matching, by definition the downconverted photons have an ordinary index of refraction, and we need to tune the effective index of refraction of the pump photons so that the signal and idler beams form a laboratory angle θ_L with the pump beam. We use equation (4.9) to calculate the refractive index $n_s = n_o$ for the downconverted photons and use Snell's law $\sin\theta_L = n_s \sin\theta_s$ to obtain θ_s. Now n_s and θ_s are inserted into equation (4.8) to determine $n_p = n_{\text{eff}}(\theta_m)$ (see figure 4.12 for the tuned index of refraction), which is used to calculate the phase matching angle θ_m using equations (4.9) and (4.10). In our case θ_m comes out to be 29.24° for $\theta_L = 3°$. Therefore, the crystals purchased were cut for a phase matching angle of 30°. Subsequently, we use kinematic mount to hold the mounted crystals so that fine-tuning of the phase-match angle can be done by adjusting the tilt of the BBO crystals.

4.2.2 Optical alignment

The optical setup and the alignment process requires care but once it has been accomplished, it is quite robust and flexible. For instance, if we are not using the downconversion crystal, we can conveniently remove it. Inserting the crystal again and tweaking its mount to fine-tune the alignment does not take a lot of time.

If one is starting from scratch, one should affix the major components to the optical bench for rough alignment, as shown in figure 4.13. First, the downconversion crystal and the collection optics of detector A are aligned. The detector collection optics include an assembly of coupling lens and a fibre optic cable. We mount each collection lens on a kinematic mount, which enables horizontal, vertical and angular adjustments. The lens height can be modified using a post holder and a post. Details of the alignment procedure are now presented.

Aligning the downconversion crystal

For the crystal alignment, the pump laser should be horizontally or vertically polarized and collimated. The laser we use is already collimated and vertically polarized. Using two mirrors, the pump beam can be aligned so that it travels level to the optical table or breadboard at a consistent height. Once aligned, the pump beam is used as a reference to align all other optical objects at the same height.

The 405 nm HWP and the downconversion crystal are inserted in the path of the pump beam. The HWP axis should be vertical.

For the sake of simplicity, we use the term 'detector' to collectively denote each set of the collection optics and a photodetector. The alignment laser is used to roughly align the detector A (more details in the next section). Once detector A is coarsely aligned, we remove the alignment laser and the connect the detector A with

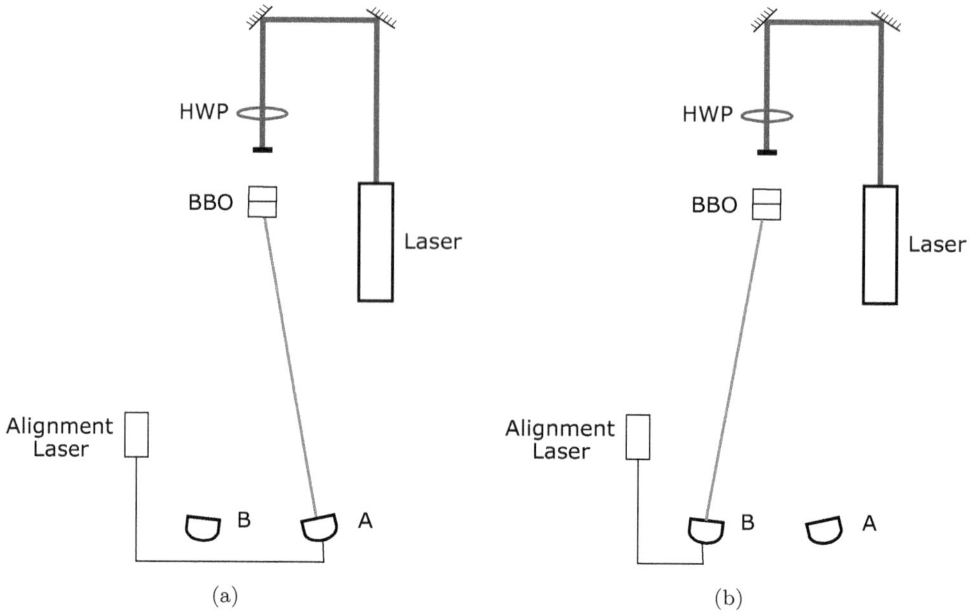

Figure 4.13. Preliminary alignment of detectors *A* and *B*. The pump beam is blocked and a visible wavelength alignment laser is back-propagated through the collection optics of (a) detector *A* and (b) detector *B* one by one. The task is to adjust the kinematic mount of each detector to impinge the alignment beam at the center of the BBO crystals.

the respective APD. Turning the lights off and turning the APDs on, the knobs of the detector *A* mount are adjusted to optimize the vertical and horizontal tilts so that the detected count rate is maximized. We slightly slide the detector *A* mount sideways to optimize the beam detection angle.

The adjustment screws on the crystal mount are used to fine-tune the tilt angles of the crystal. After aligning both the detectors, the crystal tilt is carefully adjusted again for maximizing the coincidence detection rate.

Aligning the detectors A *and* B

Since the downconverted photon beams make an angle of $\pm 3^\circ$ with reference to the pump beam, the detector *A* is placed facing the downconversion crystal, roughly at 3° with respect to the pump beam. For coarse alignment, a back-propagation laser is shone backward through the detector *A* mount and aligned towards the downconversion crystal (figure 4.13(a)). This beam actually shows where the detector *A* is pointed at. The knobs of the detector *A* kinematic mount are tweaked until the beam falls at the center of the downconversion crystal. Once it is achieved, the detector *A* has been coarsely aligned.

For finer alignment of detector *A*, we remove the alignment laser and then connect the detector *A* lens with the detector *A* APD. Turning off the lights and turning on the laser and APDs, the photon counts are monitored and the knobs of the detector *A* are adjusted such that the its counts are maximized.

Once the detector A is aligned, the next step is to align detector B such that the coincidence counts AB are maximized. The detector B is placed at a spot so that it roughly makes the same angle as does detector A, but on opposite side of the pump beam. The alignment laser is used for coarse alignment (figure 4.13(b)). Fine alignment is done with detectors connected. The knobs of detector B are adjusted such that the coincidence counts AB are maximized. It should be ensured that maximized AB counts are way above the accidental coincidence counts which are discussed in the next section.

The alignment procedure may need to be repeated a few times to obtain desirable results.

4.2.3 Accidental coincidence counts

To ensure reasonable alignment and ascertain the statistical significance of the acquired data, we need to estimate the accidental single counts and coincidence counts, of which there are three main sources. Firstly, there will be accidental detection of photons due to dark counts of the detector and detection of scattered light, which does not originate from the downconversion. Secondly, because of the finite width of coincidence time window and detector inefficiency, some photons which are not part of the same downconversion pair will be coincidentally detected, and hence give rise to accidental coincidence counts. Lastly, the coincidentally detected photons may be non-twins because of the alignment mismatch between the two detectors. All in all, these sources of accidental counts are assumed to be random.

We can model the single photon detection count rate for, say, detector A by [36]

$$N_A = N_A^t + N_A^{(\mathrm{bg})}, \tag{4.11}$$

where N_A^t is the detection count rate of true single photons produced by down-conversion and $N_A^{(\mathrm{bg})}$ is the count rate of background photons. The latter accounts for both the dark counts and the detection of scattered photons. Detector B's count rate can be modeled likewise.

Extending the argument for coincidence detections, the coincidence detection rate for both detectors A and B is given by

$$N_{AB} = N_{AB}^t + N_{AB}^{(\mathrm{acc})}, \tag{4.12}$$

where N_{AB}^t is the count rate of true coincidence events and $N_{AB}^{(\mathrm{acc})}$ is the count rate of accidental two-fold coincidences, a good approximation for which is given by [36]

$$N_{AB}^{(acc)} \approx N_A N_B \Delta t, \tag{4.13}$$

where Δt is the *effective* coincidence detection time window, which is nominally equal to 2τ where τ is the width of one pulse corresponding to one detection. Equation (4.13) treats the detections at the two detectors as independent, uncorrelated events and accounts for the possibility of their occurring within the coincidence window. Note that throughout this book, we will report all count rates as counts detected per second and, for brevity, will refer to them as simply counts.

Since we will not use the single photon counts in most cases, we will not compensate for accidental detections in single photons. However, in all our experiments, we will calculate and subtract the accidental coincidence counts from the measured coincidence counts.

4.2.4 The experiment

Experiment Q1 is essentially an exercise to properly couple the downconverted photon beams into the detector optical fibres and maximize the coincidence counts of the two beams. Aligning the two detectors, as already explained, is the major component of this experiment. Aligning the detectors actually refers to aligning the collection optics, i.e. the lens and the optical components which bring the downconverted photons to the actual photodetectors. This may look like a tedious task but this milestone, once achieved, will be the stepping stone for our subsequent experiments.

The schematic layout can be seen in figure 4.14 whereas a photograph of the experimental setup is shown in figure 4.15. Light from the pump laser, which is vertically polarized, is reflected off two mirrors and passed through an HWP before reaching the downconversion crystals. The downconverted photons fall at the collection optics of detectors A and B. The detectors output an electric pulse for each photon that they detect. These single-photon pulses as well as the pulses coincident on the two detectors are counted by the CCU. These count rates are transmitted from the CCU to the computer where they are visualized and monitored in real-time.

To see if the crystals are well-aligned, the HWP is rotated and for a range of orientations of the HWP, counts registered by the detectors A and B and coincidence counts are recorded. Since, we are using two BBO crystals with orthogonal orientations, photons of both horizontal and vertical polarization states are downconverted. Hence, for proper alignment of the crystals and detectors, we should observe minimal or no change in coincidence counts on rotating the HWP.

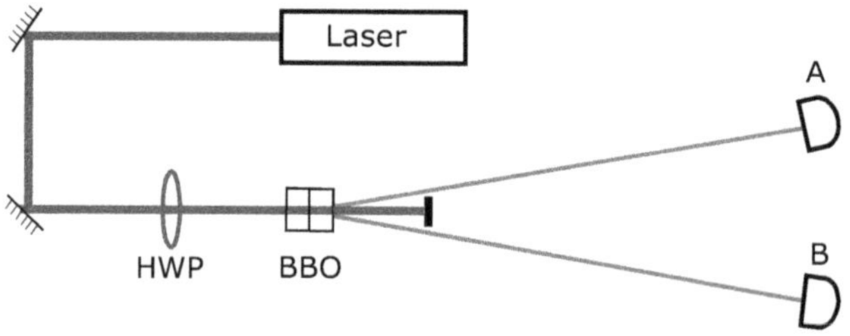

Figure 4.14. Schematic of the experimental setup for the downconversion experiment. Vertically polarized light from a violet–blue laser is reflected off two mirrors and passed through a half-wave plate (HWP) and a BBO crystal pair. Most of the light passes straight through and is blocked by a beam dump. Very few photons undergo downconversion but are efficiently picked by detectors A and B.

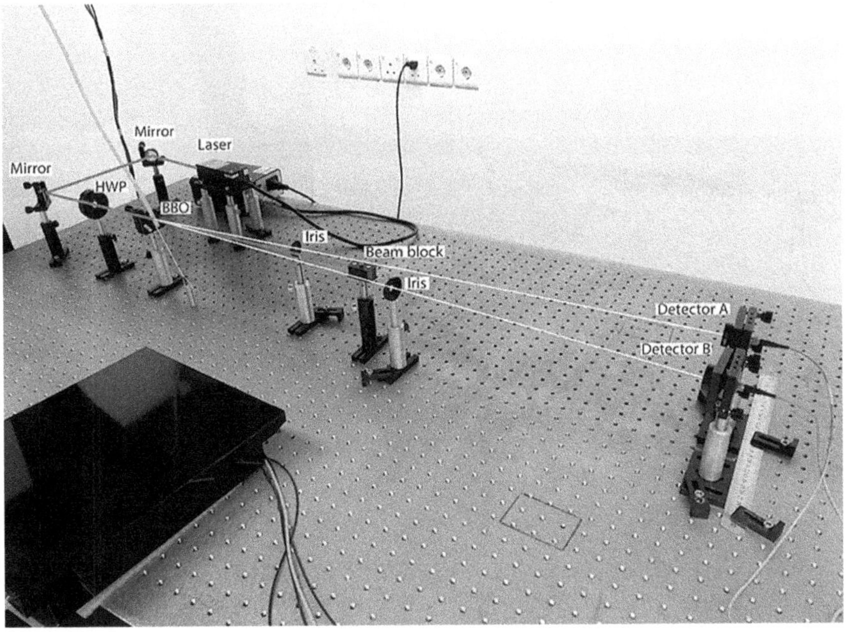

Figure 4.15. Photograph of the downconversion experiment setup for experiment Q1. The optics are mounted with screws to the table so that the alignment is robust. The irises are needed during alignment. The single-photon detectors are covered with a black case to protect them from ambient light of, say, the PC. For additional safety, the experiment is performed with room lights turned off.

Figure 4.16 shows the plot of single and coincidence counts for different orientations between 0 and π of the HWP. The periodic change of the single and coincidence counts against the change in the of the HWP implies that one of the two BBO crystals is still un-aligned. However, the three plots corresponding to detector A counts, detector B counts and AB coincidence counts are all in phase, confirming the SPDC process.

If one were to refine the alignment procedure for the downconversion crystal and then recorded the single and coincidence counts against the rotation of the HWP, the single and the coincidence counts show very little change, indicating that the BBO crystals and the detectors were well-aligned. This is illustrated in figure 4.17 and is an important benchmark towards the further experiments discussed in the rest of the report.

4.3 Q2: proof of existence of photons

Our experiments are aimed at investigating the properties of light, therefore before attempting to look at entanglement and nonlocal behavior of the quantum world, we should first verify the quantization of light. Usually, in textbooks, the quantum nature of light is introduced through a discussion of the photoelectric effect. Even though Einstein's explanation of this effect is elegant and simple, there are semiclassical theories that can explain the photoelectric effect which treat light as

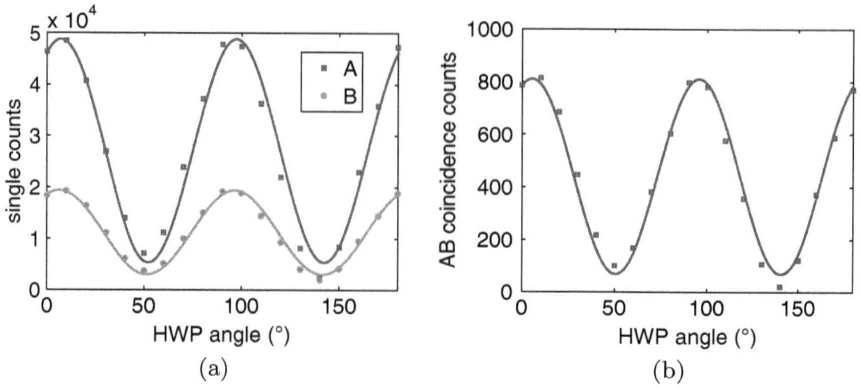

Figure 4.16. Variation of (a) single counts and (b) coincidence counts with changing orientation of the pump beam half-wave plate (HWP) prior to fine alignment. All the counts are in phase and show a sinusoidal variation, as depicted by the curve fits.

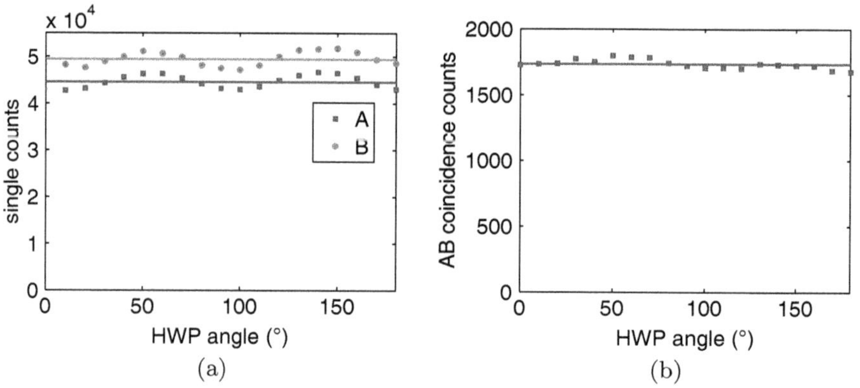

Figure 4.17. Almost no variation in (a) single counts and (b) coincidence counts with changing orientation of the pump beam half-wave plate (HWP) after the alignment has been refined. The horizontal lines show the respective mean values of the photocounts.

an electromagnetic wave and quantize the detector atoms. Such theories [37, 38] have been proposed as early as 1927.

In contrast, an experimental test to establish the existence of single photons should show the 'grainy-ness' of individual photons, i.e. it should be necessary to treat the light field quantum mechanically. In other words, the experiments should not be explainable using classical wave theory of light. On the other hand, we also like the experiment to be easily realizable. For example, certain experiments [39–41] in the 1970s demonstrated the existence of a field that would require a quantum mechanical explanation. Even though many such experiments have been subsequently performed, very few are actually realizable in an instructional laboratory [20, 33, 42].

The first experiments [43, 44] to study these intensity correlations were performed by Hanbury Brown and Twiss in 1956. Originally motivated by astronomical interferometry, their experiments revolved around measurements of second-order (intensity) correlations in light. The measured correlations were found to be higher than the expected [43, 44]. Even though the semiclassical theory accurately predicted these results, this work inspired a theoretical investigation of the statistical properties and coherence of light using purely quantum mechanical descriptions [45–47]. These studies, along with others [48–50], ultimately laid the groundwork of modern quantum optics. In this quantum 'movement', many experiments to study the statistical properties of various light sources were carried out [39, 40, 51, 52], resulting in the observation of 'anti-bunching' of photons. Anti-bunching implies coincidence measurements less numerous than the random/accidental measurements. These outcomes should not be explicable using any classical theory [41].

In this vein, an elegant and conceptually simple experiment was performed by Grangier *et al* in 1986 [28, 53]. The idea was to study correlations between photodetections at the transmission and reflection sides of an ordinary 50:50 beam splitter. If the beam consisted of individual photons, just one of the detectors (either transmission or reflection) would register a count at an instant and there would be no coincident detections. The experiment showed that only a single-photon state could account for the field incident on the beam splitter [28]. The key challenge in such an experiment, however, is *creating* a single-photon state. For this purpose, a weak laser beam containing very few photons is not enough since its statistical properties do not obey the anti-bunching prediction, even though the photon instances may be extremely sparse.

In this book, we describe an updated version of the same experiment, closely following the work described in the references [3, 25, 28]. In order to ensure the incidence of single photons on the beam splitter, we use a downconversion based heralded photon source which was described in experiment Q1. As the measurements of signal beam photons are conditioned on the photodetections of idler beam at a certain detector, the field of the signal beam effectively becomes a single-photon state.

We quantify the photodetection statistics by employing a parameter known as the degree of second order coherence, $g^{(2)}(0)$. In our experiment, we will demonstrate that if we consider the classical, wavy nature of light, we end up with the prediction $g^{(2)}(0) \geqslant 1$. Classical inequalities such as these are violated by quantum mechanics and in such cases we can claim that we are observing strictly quantum physical phenomena. We will study more inequalities in chapter 5, which deal with another interesting aspect called 'nonlocality'.

For the experiment under discussion, if our measurements result in a value of $g^{(2)}(0)$ which is less than 1, we can pronounce that classical theories do not suffice to explain the optical field. This will be considered as experimental proof of the granular nature of light. The quantum state that would show a maximal violation of this $g^{(2)}(0) \geqslant 1$ is indeed the single-photon state. For such a state, theory predicts $g^{(2)}(0) = 0$. Let's first elaborate on the meaning of $g^{(2)}(0)$ so that we can recognize how to measure it for our single-photon experiments.

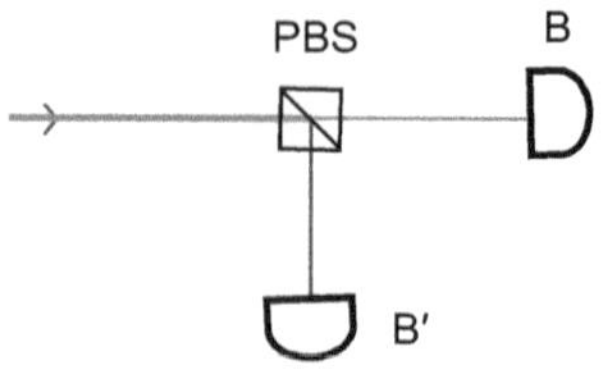

Figure 4.18. A typical polarizing beam splitter (PBS) lets the horizontally polarized component of the incident light to transmit through it and reflects the vertically polarized component. The transmitted intensity, in this case, is detected by detector B while the reflected intensity is detected by detector B'.

4.3.1 Photodetection and degree of second-order coherence

In this section, we model the photodetection process and present the classical as well as quantum predictions for $g^{(2)}(0)$.

The classical view

When we say classical field, we refer to an electromagnetic wave whose behavior is completely encompassed by Maxwell's equations (see equations (2.1)–(2.4) in chapter 2). Referring to figure 4.18, consider an optical field which is directed towards a polarizing beam splitter (PBS). Some part of this incident beam is transmitted and falls on detector B and the other part of the beam is reflected and falls on detector B'. Let $I_I(t)$ denote the intensity of the input field and let $I_B(t)$ and $I_{B'}(t)$ represent the respective intensities of the beams falling on the detectors. Then, for gauging the correlations between the detector intensities $I_B(t)$ and $I_{B'}(t)$, we can use the second-order coherence function. This is denoted as $g^{(2)}_{B,B'}(\tau)$ and expressed as [3, 54]:

$$g^{(2)}_{B,B'}(\tau) = \frac{\langle I_B(t+\tau) I_{B'}(t)\rangle}{\langle I_B(t+\tau)\rangle\langle I_{B'}(t)\rangle}, \tag{4.14}$$

where τ represents the time delay between the two intensity measurements. We call the quantity in equation (4.14) the degree of second-order coherence since it is based on intensity correlations. In contrast, the degree of first-order coherence is used for a function involving correlations between fields [3].

We are interested in intensity correlations that are simultaneous, i.e. we would like to study $g^{(2)}_{B,B'}(\tau)$ at $\tau = 0$. If we represent the transmission coefficient of the splitter as T, and the reflection coefficient as $R = 1 - T$, it is straightforward to see that the transmitted and reflected intensities are given by $I_B(t) = TI_I(t)$ and $I_{B'}(t) = RI_I(t)$. Inserting these expressions into equation (4.14) gives

$$g^{(2)}(0) = g^{(2)}_{B,B'}(0) = \frac{\langle [I_I(t)]^2\rangle}{\langle I(t)\rangle^2}. \tag{4.15}$$

From the Cauchy–Schwarz inequality, $\langle [I_I(t)]^2\rangle \geqslant \langle I(t)\rangle^2$, and hence, for classical fields, the equation above yields

$$g^{(2)}(0) \geqslant 1. \tag{4.16}$$

In equation (4.16), equality with one (1) is obtained if there are no fluctuations in the input intensity. In contrast, if we have a fluctuating input field intensity, $g^{(2)}(0)$ is found to be greater than 1. The best approximation for a stable input source that we can experimentally achieve is monochromatic, ultra-stable laser light, which yields $g^{(2)}(0) = 1$. On the other hand, in case of 'chaotic' light, one can demonstrate that $g^{(2)}(0) = 2$ [54]. An example of such a source is light received from a vapor lamp.

Now that we have discussed correlations between light intensities exiting the beam splitter, we need to model the photodetection process. This is important because while doing an experiment, we do not perform a direct measurement of the intensity but instead we measure the output current from a photodetector.

The semiclassical view of photodetection

According to the semiclassical theory of photoelectric detection [3], photoelectrons are randomly generated due to incident electromagnetic waves. Using a photodetector (say detector B), the probability of registering one photocount within a short interval represented by Δt is given by

$$P_B = \eta_B \langle I_B(t) \rangle \Delta t, \tag{4.17}$$

where $\langle I_B(t) \rangle$ represents the average intensity of the field incident on the detector and η_B denotes the photodetection efficiency. Likewise, the joint probability of registering a photocount (in a short window Δt) at detector B', and then for obtaining a photocount at detector B after time τ (also within an identical time window Δt), is given by

$$P_{BB'}(\tau) = \eta_B \eta_{B'} \langle I_B(t + \tau) I_{B'}(t) \rangle (\Delta t)^2. \tag{4.18}$$

From equations (4.14), (4.17), and (4.18), we obtain

$$g^{(2)}_{BB'}(\tau) = \frac{P_{BB'}(\tau)}{P_B P_{B'}}, \tag{4.19}$$

meaning that the degree of second-order coherence can be determined by measuring the probability of single and coincident photodetections at detectors B and B'. For $\tau = 0$ and using equations (4.19) and (4.16), we find once again that for classical fields,

$$g^{(2)}_{BB'}(0) = \frac{P_{BB'}(0)}{P_B P_{B'}} = g^{(2)}(0) \geqslant 1. \tag{4.20}$$

The probability of registering a photodetection at detector B within a small interval Δt can be simply computed by multiplying the average photodetection rate with Δt. To compute the average photodetection rate for detector B, we simply need to divide the number of photodetections N_B by the total time T over which we are counting, where T is sometimes called the integration time. We can determine the probabilities for individual detections at detector B' ($N_{B'}$) and BB' coincidence detections ($N_{BB'}$) in a similar fashion:

$$P_B = \left(\frac{N_B}{T}\right)\Delta t, \tag{4.21a}$$

$$P_{B'} = \left(\frac{N_{B'}}{T}\right)\Delta t, \tag{4.21b}$$

$$P_{BB'} = \left(\frac{N_{BB'}}{T}\right)\Delta t. \tag{4.21c}$$

In our case, the variable Δt corresponds to the coincidence time window of the CCU. Substituting equation (4.21) into equation (4.20), we obtain

$$g^{(2)}(0) = \frac{N_{BB'}}{N_B N_{B'}}\left(\frac{T}{\Delta t}\right). \tag{4.22}$$

This expression is related to the two-detector measurement of $g^{(2)}(0)$, exemplified in figure 4.18, and hence it must not be confused with the three-detector version which we will discuss in the following section.

The quantum view

In a three-detector experiment, we still measure $g^{(2)}(0)$ for a beam incident on a beam splitter. The difference is that the photodetection of this beam is now conditioned on the photodetection of another beam. The conditioning is crucial—photodetection of the idler beam ensures the single-photon state of the signal beam. Otherwise, the optical beam falling on the beam splitter will be classical. For a comparison, when doing the three-detector measurement, we will also execute an unconditional measurement on the signal beam. The latter will form a two-detector measurement, and we will observe $g^{(2)}(0) \geqslant 1$ in this case. Let us now discuss the three-detector experiment.

Consider the measurement of $g^{(2)}(0)$ using the three-detector setup and when all the probabilities are conditioned on the photodetections at a third detector, which is detector A. The scheme is shown in figure 4.19. For this experimental setup, instead of equation (4.20), we have the following expression for $g^{(2)}(\tau)$ at $\tau = 0$:

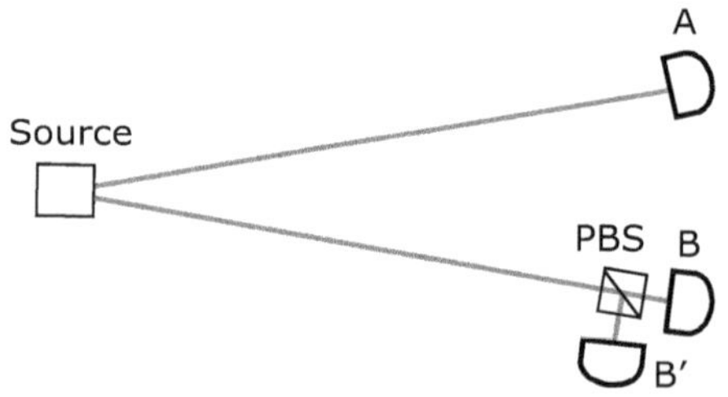

Figure 4.19. The source produces pairs of photons, one of which goes to detector A and the other goes to the polarizing beam splitter (PBS), with two detectors B and B' placed at the transmission and reflection outputs of the PBS. Detection of a photon at detector A projects the other photon into a single-photon state.

$$g^{(2)}(0) = \frac{P_{ABB'}(0)}{P_{AB}(0)P_{AB'}(0)}, \tag{4.23}$$

where $P_{ABB'}(0)$ represents the probability of threefold coincidence event involving all the three detectors A, B and B', and $P_{AB}(0)$ and $P_{AB'}(0)$ represent the probabilities of two-fold coincidence events at the detectors A, B and B'.

We want to monitor only the events in which detector A is triggered. Hence we can take the number of photodetections at this detector and use it as the total number of trials. Representing this number as N_A, we use it to normalize the photodetection probabilities as follows:

$$P_{AB}(0) = \frac{N_{AB}}{N_A}, \tag{4.24a}$$

$$P_{AB'}(0) = \frac{N_{AB'}}{N_A}, \tag{4.24b}$$

$$P_{ABB'}(0) = \frac{N_{ABB'}}{N_A}. \tag{4.24c}$$

Plugging these probabilities into the $g^{(2)}(0)$ expression finally yields for the degree of second-order coherence,

$$g^{(2)}(0) = \frac{N_A N_{ABB'}}{N_{AB} N_{AB'}}. \tag{4.25}$$

If an individual photon is indeed incident on the PBS in figure 4.19, we should get $N_{BB'} = 0$ in equation (4.22), and $N_{ABB'} = 0$ in equation (4.25). For both the two-detector and the three-detector setups, the quantum mechanical prediction is therefore $g^{(2)}(0) = 0$ if the incident field comprises single photons.

A classical light source is bound to satisfy the inequality $g^{(2)}(0) \geqslant 1$. In the case that the photodetections being monitored at detectors B and B' are totally uncorrelated, we obtain $g^{(2)}(0) = 1$. This happens if a completely stable light beam falls on the beam splitter. Such a beam is exemplified by an ultra-pure, ultra-stable laser. In contrast, when the beam falling on the beam splitter is characterized by fluctuations, we obtain $g^{(2)}(0) > 1$, which implies detection of positive correlations between the photodetection events. We observe such fluctuations in a thermal light source. Hence, for a classical light source, we cannot have $g^{(2)}(0)$ smaller than 1. This implies that in the case of classical fields, we cannot have anti-correlated photodetections at the detectors placed at the two outputs of a beam splitter.

This argument can also be intuitively understood. The function of a beam splitter is to simply divide an optical field into two daughter fields. These resultant fields may either swing together or may not swing at all. In the former case, there is a positive correlation in the fluctuations of the field, which is called 'bunching'. In the latter case there are not fluctuations and hence no correlations. Therefore classically, it does not make sense if during the fluctuations, there is an increase in intensity of one of the daughter fields while at the same time there is a decrease in the intensity of the other daughter field. Such a situation would imply anti-correlations and is termed as 'anti-bunching'.

We would like to point to a subtlety involved in this experiment. To ensure a measurement approaching $g^{(2)}(0) = 0$, we are required to ensure single-photon incidence on the PBS. By using our SPDC source, this requirement is achieved by conditioning the signal photon detection to an idler photon at detector A. Therefore, for such a light source, the requirement of a three-detector setup is crucial to obtain $g^{(2)}(0) = 0$. If we employ the downconversion source and do not perform the conditioning of detection, we will not achieve single-photon incidence on the PBS. Measurements with this setup will consequently result in $g^{(2)}(0) \geqslant 1$. In our experiments, we use the same setup for two-detector and three-detector $g^{(2)}(0)$ measurements but look at different sets of single and coincidence counts, as required by equations (4.22) and (4.25) respectively, for the two schemes.

Finally, while performing an experiment in the laboratory, we should not expect ideal measurements that would yield $g^{(2)}(0) = 0$. This is due to the presence of accidental coincidences. We can predict these accidental coincidences using the coincidence time window Δt, the integration time T, and the average values of the single detector counts. Using these predicted accidental coincidence detections, the expected $g^{(2)}(0)$ can be determined.

Another way to look at the statistics of photodetection to classify light sources is as follows (figure 4.20) [54]. A stable light source with constant intensity such as coherent light from a laser, gives Poissonian statistics ($\Delta n = \sqrt{\bar{n}}$). If there are any classical fluctuations in the intensity, larger photon number fluctuations are expected as compared to the case with a constant intensity. All classical light beams of this kind with time-varying light intensities show super-Poissonian $\Delta n > \sqrt{\bar{n}}$ photon

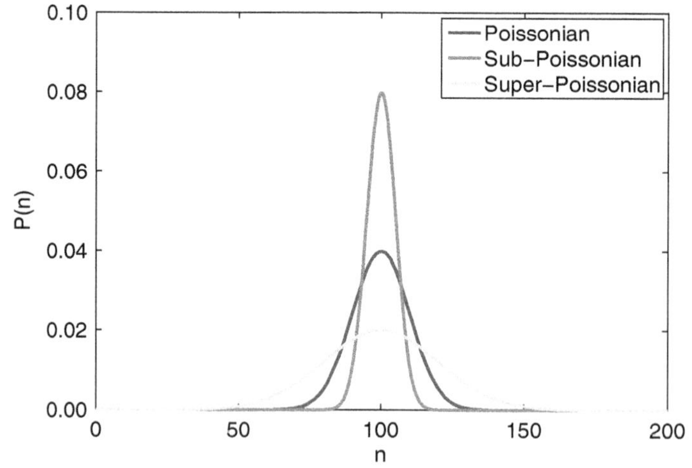

Figure 4.20. Comparison of Poissonian, sub-Poissonian and super-Poissonian probability distributions for 100 mean photodetections ($\bar{n} = 100$). The sub-Poissonian distribution has the smallest spread while the super-Poissonian distribution has the largest.

number distributions. Sub-Poissonian statistics $\Delta n < \sqrt{\bar{n}}$, by contrast, have a narrower distribution than the Poissonian case, and is indeed the case when detecting single photons.

4.3.2 Accidental coincidences

The time window for coincidence detection is constrained by the internal clock of the FPGA, which implements the CCU. Consequently, this finite time interval represented as Δt results in a finite probability of detecting accidental ABB' coincidences. This probability is proportional to Δt. For instance, within Δt of any valid coincidence detection given by P_{AB}, there is a random chance that the B' detector also will register a count, and this results in an accidental threefold coincidence. We can express this probability of accidental coincidence detection as

$$P'_{ABB'} = P_{AB}P'_{B'} + P_{AB'}P'_{B} = P_{AB}N_{B'}\Delta t + P_{AB'}N_{B}\Delta t. \tag{4.26}$$

It is safe to ignore the probability of accidental threefold coincidence detection which is the result of purely random detections simultaneously occurring at the three detectors. This is because such a probability is already negligible for our count rates and coincidence window. In order to determine the propagation of error in the correlation function due to the accidental coincidence counts, we substitute equation (4.26) into equation (4.23), which results in

$$\begin{aligned} g^{(2)}(0)' &= \frac{P'_{ABB'}}{P_{AB}P_{AB'}} \\ &= \frac{P_{AB}N_{B'}\Delta t + P_{AB'}N_{B}\Delta t}{P_{AB}P_{AB'}} \\ &= \frac{N_{B'}\Delta t}{P_{AB'}} + \frac{N_{B}\Delta t}{P_{AB}} \\ &= N_{A}\Delta t\left(\frac{N_{B'}}{N_{AB'}} + \frac{N_{B}}{N_{AB}}\right), \end{aligned} \tag{4.27}$$

showing that the average counts obtained during the experiment can be used to compute the contribution of accidental coincidence events to $g^{(2)}(0)$. We subtract these accidental counts from the experimental counts received as described in later sections.

4.3.3 The optical setup

In experiment Q1, we examined the behavior of an SPDC source. Our attempts were aimed at maximizing the coincidence detection rate between two detectors, namely A and B. Experiment Q2 builds up on the same setup, where detectors A and B have been already aligned. The major task in this experiment is inserting a PBS in the path of signal beam, and then aligning the detector B'. As soon as this is achieved, we can measure $g^{(2)}(0)$.

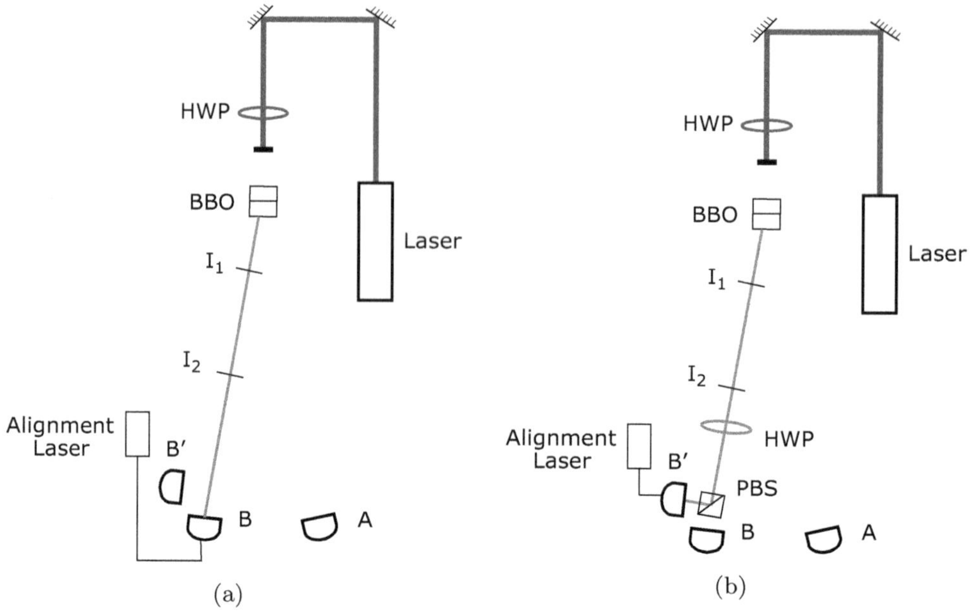

Figure 4.21. Rough alignment of detectors B and B' with a polarizing beam splitter (PBS) placed in the path. The pump beam is blocked and a visible wavelength alignment laser is back-propagated through the collection optics of (a) detector B and (b) detector B' one by one. Two irises I_1 and I_2 are used to assist the alignment procedure and center the alignment beam at the center of the BBO crystals.

The alignment laser is shone backwards through the detector B. If the laser beam does not point towards the center of the downconversion crystal, we need to perform alignment of detector B (following the method of experiment Q1) before continuing.

Furthermore, two irises are placed in the path of the alignment beam (figure 4.21(a)). One should be close to the detector, and the other should be close to the downconversion crystal.

The PBS is inserted just about a distance of 3–4 inches from the detector B (figure 4.21(b)). It must be ensured that the face of the PBS is perpendicular to the laser beam and that the beam passes through its center. We ensure that the PBS is placed in a way that the light traveling from the source is reflected away from the beam falling on detector A.

Next, an HWP is inserted between the PBS and the downconversion crystal, close to the PBS. The procedure for aligning the B' detector in a more refined manner is detailed as follows.

Aligning the B′ *detector*

The second major task is alignment of the B' detector with the aim to make sure that both detectors B' and B receive light from the same downconverted beam. The collection optics of the detector B' is placed at the reflection end of the beam splitter. It should be ensured that the both detectors B' and B are equidistant from the beam splitter. Now the alignment laser is back-propagated through the B' detector, as shown in figure 4.21(b). We need to ensure that the alignment beam passes successfully through both irises and ultimately hits the center of the downconversion crystal.

We do this by sliding the detector mount back and forth, and sideways in order to make the beam traverse through the center of iris I_2. Using the knobs on the mount, its tilt is tweaked so that the beam is centered on the iris I_1. We perform a few iterations of these two adjustments. Once detector B' is well-aligned to a good extent, it can be screwed to the optical bench or breadboard. Now, once again the vertical and horizontal tilts of the B' collection optics are adjusted so that the beam is perfectly centered on iris I_2. Next, the vertical and horizontal tilts of the beam splitter are tweaked to center the beam on iris I_1. When the beam is well-centered on both the irises, we remove the alignment laser and connect the B' detector with the SPCM B'. Turning off the lights and turning on the pump laser and the detectors, we open the irises wide and start monitoring the count rates.

The HWP placed in one of the downconverted beam is carefully rotated while the count rates are monitored. For some HWP orientations, there will be a large number of B and AB detections, but at the same time, a negligibly small number of B' and AB' detections. Likewise, for some other orientations of the HWP, there will be maximal B' and AB' detections, but very few B and AB detections. At this stage, the HWP is rotated in order to achieve maximum number of AB detections and then the tilt of the B detector mount is adjusted to achieve further maximization of the AB coincidence detections. Subsequently, a similar technique is used for the AB' counts. The HWP is rotated and the tilt on the B' mount is adjusted to achieve maximum detections for the AB' coincidence events.

To optimally perform the experiment, it is recommended that there should not be a significant difference between the maximum values for AB and AB' coincidence detections. If the counts are drastically different, alignment may need to be improved or re-done. The optic fibres for one of the B and B' detectors may also need cleaning, or it might be the case that the fibre coupling lens for one of the detectors may have a better alignment than that of the other.

4.3.4 The experiment

The laser which is vertically polarized is reflected off two mirrors, passes through an HWP oriented vertically, and is downconverted by the pair of mutually stacked BBO crystals. The laser and the detectors A and B are aligned, based on the procedure of experiment Q1. In the detector B path, a PBS is introduced. Another detector B' is placed and aligned in front of the reflection side of the PBS. The schematic layout is shown in figure 4.22.

Counts registered at detectors A, B and B', and coincidence counts AB, AB', BB' and ABB' are recorded. AB and AB' counts are plotted in figure 4.23(a) for different orientations of the HWP. In figure 4.23(a), the AB and AB' counts vary sinusoidally and in fact correspond to the downconverted photons projected onto $|HH\rangle$ and $|VV\rangle$ respectively as described briefly in experiment Q1. Their dependences on the half-wave plate angle θ_{HWP} are respectively, $\cos^2(2\theta_{\text{HWP}})$ and $\sin^2(2\theta_{\text{HWP}})$. For

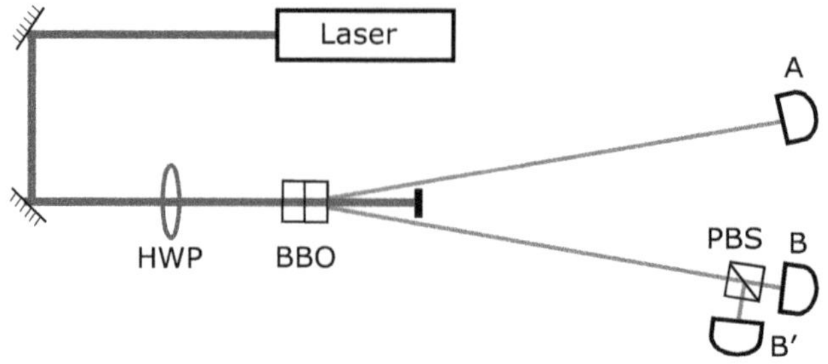

Figure 4.22. Vertically polarized light from a violet–blue laser is reflected off two mirrors and passed through a half-wave plate (HWP) and BBO crystals. The downconverted idler photons go to detector A and the signal photons travel towards the polarizing beam splitter (PBS), having transmission end towards detector B and reflection end towards detector B'.

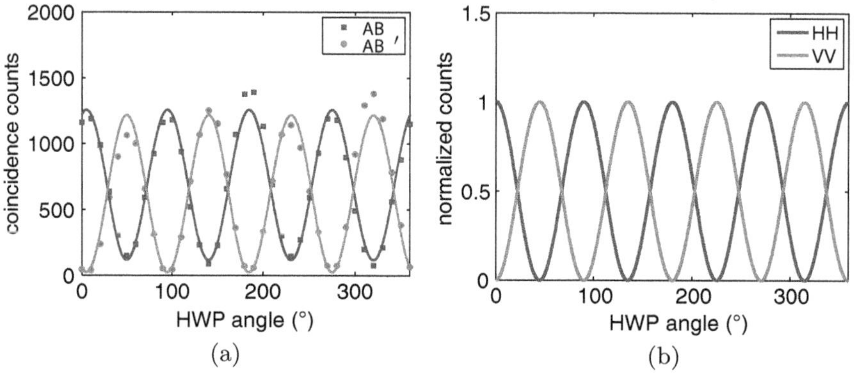

Figure 4.23. Variation of AB and AB' coincidence counts with respect to change in orientation of the pump beam half-wave plate (HWP). (a) Measured coincidence counts with curve fits $1139\cos^2(2\theta_{\rm HWP} - 10) + 117$ and $1191\sin^2(2\theta_{\rm HWP} - 10) + 25$ (where $\theta_{\rm HWP}$ is in degrees), corresponding to AB and AB' respectively. (b) Simulated coincidence counts where HH represents the photon pairs predicted in the state $|HH\rangle$, corresponding to AB; and VV represents the photon pairs predicted in the state $|VV\rangle$, corresponding to AB'.

comparison, simulated normalized counts are also plotted in figure 4.23(b). The experimental and the simulated results agree very well and let us say the following confidently. The PBS lets the horizontally polarized photons pass through it and reflects the vertically polarized photons. The horizontally polarized photons are detected at detector B while the vertically polarized photons are detected at detector B'. Therefore when we look at AB coincidences, we are essentially counting $|HH\rangle$ photon pairs and when we look at AB' coincidences, we are actually counting $|VV\rangle$ photon pairs. While counting $|HH\rangle$ photon pairs, $|V\rangle$ detections at detector A do not count towards coincidences because of no simultaneous detections at detector B and vice versa.

For the same setup of figure 4.22, $g^{(2)}(0)$ is determined for two-detector and three-detector schemes for different pump beam HWP orientations. The two-detector scheme involves just B and B' detectors while the three-detector one includes detector A as well. See the $g^{(2)}(0)$ plots in figure 4.24. It is quite evident that for

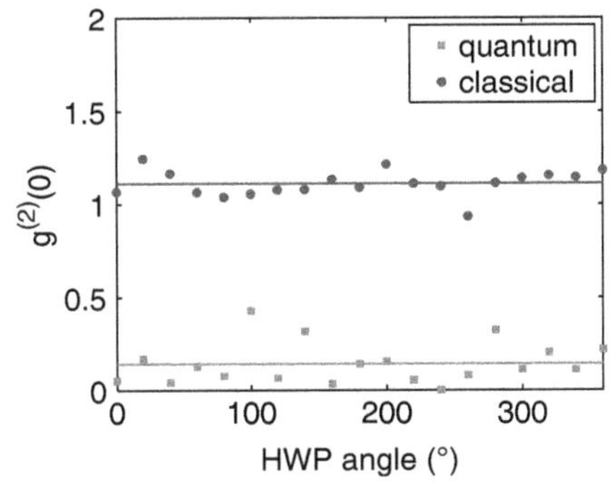

Figure 4.24. Two-detector (classical) and three-detector (quantum) degree of second order coherence $g^{(2)}(0)$ for various orientations of the half-wave plate (HWP). The smooth lines, corresponding to the respective mean values, act as guide to the eye.

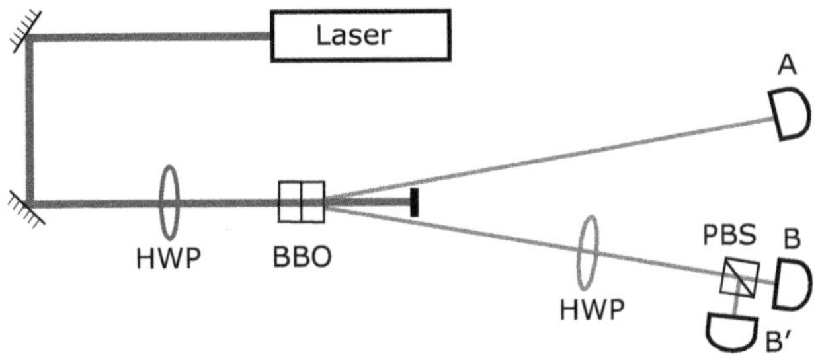

Figure 4.25. Schematic diagram of the experimental setup to measure three-detector $g^{(2)}(0)$. The pump beam half-wave plate (HWP) is oriented at 0° with respect to the pump beam polarization and the second HWP is oriented such that the *AB* and *AB*′ counts are approximately equal.

the two-detector setup, $g^{(2)}(0)$ fluctuates around one throughout the rotation of the HWP whereas for the three-detector scheme, it stays close to zero. This means that in the former case, $g^{(2)}(0)$ values show photodetection of coherent classical beam of light whereas the latter case points toward detection of single photons, which implies that detection events at detector *A* ensure detection of single photons at detectors *B* and *B*′.

Now, the HWP is oriented at 0° with respect to the vertical and another HWP is introduced between the PBS and the downconversion crystal. The scheme is shown in figure 4.25 and is photographed in figure 4.26. This second HWP is oriented such that the *AB* and *AB*′ counts are roughly equal. Two independent datasets for the three-detector $g^{(2)}(0)$ are recorded. See table 4.1.

For perfectly single photons, we should have obtained $g^{(2)}(0) = 0$, i.e. we expect no coincidences between detectors *B* and *B*′. Experimentally, we obtained non-zero $g^{(2)}(0)$, which can be explained *solely* by the accidental coincidences, as shown in table 4.1. These results, without a shade of doubt, verify the quantum nature of light —the phenomenon of anti-bunching—and confirm the existence of single photons.

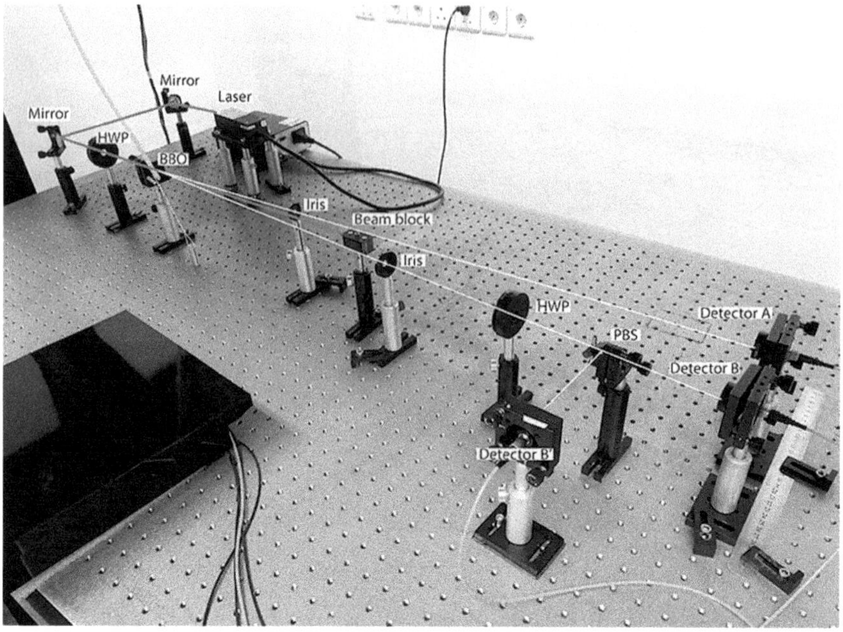

Figure 4.26. Photograph of the three-detector $g^{(2)}(0)$ measurement setup.

Table 4.1. Results of three-detector $g^{(2)}(0)$ measurements. Results for two runs are shown. The column labeled 'Confidence' lists the number of standard deviations σ by which the experimental results violate $g^{(2)}(0) \geqslant 1$.

Integration time	Experimental $g^{(2)}(0)$	Accidental $g^{(2)}(0)$	Confidence
125 s	0.111 ± 0.013	0.119	67 σ
630 s	0.079 ± 0.005	0.087	182 σ

In order to investigate $g^{(2)}(\tau)$ further, the three-detector experiment is repeated for different values of time delay τ, introduced in the detector B signal by passing it through a delay generator[4]. Figure 4.27(a) shows typical $g^{(2)}(\tau)$ for bunched, coherent and anti-bunched (quantum) light whereas figure 4.27(b) shows the experimentally determined $g^{(2)}(\tau)$ obtained for our experiment. The plot in figure 4.27(b) illustrates a drop in the value of $g^{(2)}(\tau)$ at $\tau = 0$. This implies anti-bunching of photons, i.e. the B and B' photodetections are not correlated. In other words, a single photon is detected at either detector B or B' but not at both simultaneously. This is a truly quantum signature of light and agrees with the simulated $g^{(2)}(\tau)$ for a quantum light source, shown in figure 4.27(a).

[4] A delay generator is implemented by a coax delay box (Stanford Research Systems DB64).

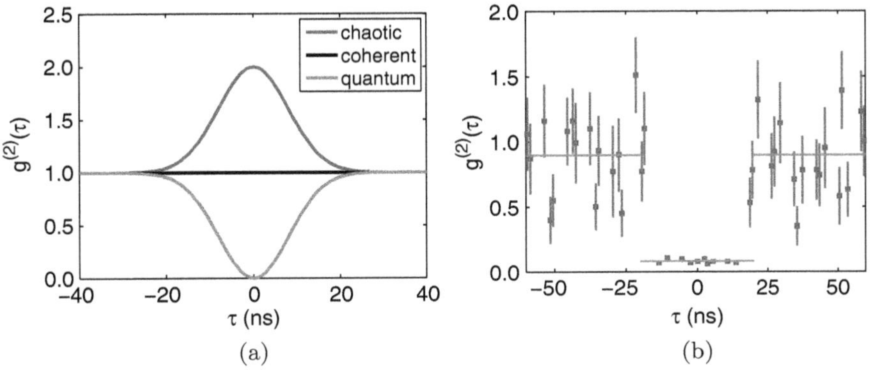

Figure 4.27. (a) Simulated $g^{(2)}(\tau)$ for chaotic, coherent and quantum light source. (b) Experimentally measured $g^{(2)}(\tau)$ for the single-photon source. The piece-wise curve fits act as guide to the eye.

Looking at figures 4.27(a) and 4.27(b), one can see that the experimental $g^{(2)}(\tau)$ stays close to zero for a longer delay τ around 0 ns. This is because two pulses that arrive within the coincidence window of our CCU are counted as coincident. So within this window, the delay generator does not create significant difference in the number of coincidence detections and hence the value of $g^{(2)}(\tau)$. This experiment is a photonic analog of the classic Hanbury Brown–Twiss experiment [43] and confirms the grainy nature of light.

4.4 Q3: Estimating the polarization state of single photons

In experiments Q1 and Q2, we were mainly concerned with the statistics of photodetection or the grainy nature of light itself. In experiments Q3 and Q4, we will study techniques for estimating the polarization state of the signal beam of downconverted photons. We will produce separable states, i.e. $|HH\rangle$ or $|VV\rangle$, through downconversion. The idler beam will be used to herald single-photon detections of the signal beam, for which we can safely assume the polarization state to be pure. Therefore, we can express the state of the signal beam as a state vector or a ket. We will create arbitrary polarization states with the signal photons and perform appropriate measurements to estimate the states.

As discussed in chapter 3, there also exists a more general representation of an arbitrary quantum state in terms of a density matrix. We postpone the experimental investigation of estimating the density matrix of an arbitrary polarization-encoded state until chapter 6. In particular, at that juncture, we will determine the complete polarization state of the downconverted photon pair in the experiment labeled QST.

4.4.1 Reconstructing the single-photon polarization state

Experiment Q3 closely follows the method presented in [3], and will be revisited from alternative perspectives in experiments Q4 and QST. Just as in the previous experiments, the idler photons will be used to herald the detection of signal photons. We are not interested in the polarization state of idler photons in this experiment and will only determine the polarization state of the signal photon.

First, we give an overview of the method to estimate the polarization state of an ensemble of identically prepared single photons. The quantum state of the signal photons encode the polarization state and, as discussed in chapter 3, can be generally represented as

$$|\Psi\rangle = A|H\rangle + Be^{i\phi}|V\rangle, \tag{4.28}$$

where $A^2 + B^2 = 1$, and A, B and ϕ are real numbers. So our task boils down to determining the three parameters A, B and ϕ. To accomplish this, we perform a large number of polarization measurements on ensembles of photons that are identically prepared in the state $|\Psi\rangle$. We use three bases $\{|H\rangle, |V\rangle\}$, $\{|D\rangle, |A\rangle\}$ and $\{|L\rangle, |R\rangle\}$ to perform these measurements so that the probabilities $P(H||\Psi\rangle)$, $P(D||\Psi\rangle)$, and $P(L||\Psi\rangle)$ can be determined. We employ these probabilities to compute the real-valued parameters describing the state $|\Psi\rangle$ to be estimated.

If we take a beam of photons identically generated in the state $|\Psi\rangle$ and perform a large number of polarization measurements in the canonical basis $\{|H\rangle, |V\rangle\}$, according to Born's rule [3], we can determine the probability of detecting a horizontally polarized photon as

$$P(H||\Psi\rangle) = |\langle H|\Psi\rangle|^2 = A^2. \tag{4.29}$$

From this probability and the normalization constraint on the state vector, we can actually obtain both A and B as

$$A = \sqrt{(P(H||\Psi\rangle)} \text{ and} \tag{4.30}$$

$$B = \sqrt{1 - A^2}. \tag{4.31}$$

Once we determine A and B, we would like to determine ϕ which requires a similar set of measurements but in the $\{|D\rangle, |A\rangle\}$ and $\{|L\rangle, |R\rangle\}$ bases. It is important that the state generation setup must not change during all these measurements. We do not want to modify the state $|\Psi\rangle$ until it has been probed by all the measurements in the three bases.

We change the measurement basis to $\{|D\rangle, |A\rangle\}$ and perform a large number of measurements. Once again, we can compute the probability of detecting a diagonally polarized photon using Born's rule as

$$\begin{aligned} P(D|||\Psi\rangle) &= |\langle D|\Psi\rangle|^2 \\ &= |\langle D|(A|H\rangle + Be^{i\phi}|V\rangle)|^2 \\ &= \frac{1 + 2AB\cos\phi}{2}. \end{aligned} \tag{4.32}$$

As we have already determined both A and B, we can invert equation (4.32) to obtain $\cos\phi$ from $P(D||\Psi\rangle)$. However, this does not uniquely determine ϕ because the inverse of cosine is not unique. Hence, measurements in a third basis must be performed on photons generated in the same state. Therefore, we now modify the apparatus to perform measurements in the $\{|L\rangle, |R\rangle\}$ basis. Now the probability of detecting a left circularly polarized photon is given by

$$\begin{aligned} P(L||\Psi\rangle) &= |\langle L|\Psi\rangle|^2 \\ &= |\langle L|(A|H\rangle + Be^{i\phi}|V\rangle)|^2 \\ &= \frac{1 + 2AB\sin\phi}{2}. \end{aligned} \tag{4.33}$$

We can manipulate this equation to obtain $\sin\phi$. Finally, knowledge of both $\cos\phi$ and $\sin\phi$ lets us uniquely determine ϕ. In fact, one can obtain the following useful expression for ϕ by manipulating equations (4.32) and (4.33) [3]:

$$\phi = \tan^{-1}\left(\frac{P(L||\Psi\rangle) - 0.5}{P(D|||\Psi\rangle) - 0.5}\right). \tag{4.34}$$

4.4.2 Generating and measuring polarization states

As in the experiments Q1 and Q2, detecting an idler photon is essential to prepare a single-photon state in the signal beam. An arbitrary polarization of the single-photon state can be generated simply by adding an HWP or a quarter-wave plate (QWP) in the path of the signal beam right after the downconversion crystal. So, if the downconversion produces horizontally polarized photons, this HWP can be used to generate any linear polarization. We borrow the transformation matrix $\hat{O}_{\text{HWP}}$ for this operation from table 3.2 and compute the generated state as

$$\begin{aligned} |\Psi'\rangle &= \hat{O}_{\text{HWP}}|H\rangle \\ &= \begin{pmatrix} \cos 2\theta & \sin 2\theta \\ \sin 2\theta & -\cos 2\theta \end{pmatrix}\begin{pmatrix} 1 \\ 0 \end{pmatrix} \\ &= \begin{pmatrix} \cos 2\theta \\ \sin 2\theta \end{pmatrix}. \end{aligned} \tag{4.35}$$

From equation (4.35), it can be shown that with HWP orientations $\theta = 0°$, $\theta = 45°$, $\theta = 22.5°$ and $\theta = -22.5°$, states $|H\rangle$, $|V\rangle$, $|D\rangle$ and $|A\rangle$ can be generated. Hence, we can label this HWP as the 'state-generation' HWP. On the other hand, to generate the circular or elliptical polarization state, a state-generation QWP is used in place of the state-generation HWP. Once again, we obtain the transformation matrix $\hat{O}_{\text{QWP}}$ from table 3.2 and compute the generated state as

$$\begin{aligned} |\Psi'\rangle &= \hat{O}_{\text{QWP}}|H\rangle \\ &= \begin{pmatrix} \cos^2\theta + i\sin^2\theta & (1 - i)\sin\theta\cos\theta \\ (1 - i)\sin\theta\cos\theta & \sin^2\theta + i\cos^2\theta \end{pmatrix}\begin{pmatrix} 1 \\ 0 \end{pmatrix} \\ &= \begin{pmatrix} \cos^2\theta + i\sin^2\theta \\ (1 - i)\sin\theta\cos\theta \end{pmatrix}. \end{aligned} \tag{4.36}$$

From equation (4.36), it can be shown that states $|L\rangle$ and $|R\rangle$ can be generated with $\theta = 45°$ and $\theta = -45°$ respectively.

Once a polarization state is generated, it needs to be measured. The measurement always refers to some basis. Intuitively, the idea of a measurement basis can be understood as follows. If we want to measure in the $\{|H\rangle, |V\rangle\}$ basis, the measuring apparatus should be able to distinguish between $|H\rangle$ and $|V\rangle$ polarized light. Measurements in the this basis can be readily performed using a PBS, which transmits horizontally polarized photons and deflects vertically polarized photons through 90° to a different trajectory. The probability that an individual photon will transmit through the PBS and thus fall at detector B is equal to the probability of registering an AB coincidence event. This probability can be computed from the measured coincidence counts as follows:

$$P_{AB} = \frac{N_{AB}}{N_{AB} + N_{AB'}}. \tag{4.37}$$

Similarly, the PBS reflects the vertically polarized photons which then travel towards detector B', and hence the probability that an individual photon is reflected off the PBS is given by

$$P_{AB'} = \frac{N_{AB'}}{N_{AB} + N_{AB'}}. \tag{4.38}$$

Now, we can perform measurements in the $\{|D\rangle, |A\rangle\}$ basis by placing an HWP oriented at 22.5° in the path of the beam just before the PBS. Similarly, we can perform measurements in the $\{|L\rangle, |R\rangle\}$ basis by placing a QWP oriented at 45° before the PBS. However, instead of plugging in and out optical components, we use the setting shown in figure 4.28 for experimental convenience. The signal beam first passes through a QWP, then an HWP, and then falls on the PBS. The PBS sends each individual photon of the beam to either of the two detectors B and B'. We just need to rotate the wave plates to change the measurement basis according to table 4.2, instead of removing or inserting any components during the experiment.

It can be shown that, in the $\{|D\rangle, |A\rangle\}$ basis according to table 4.2, P_{AB} corresponds to the probability of detecting photons in the state $|D\rangle$ while $P_{AB'}$ corresponds to the probability of detecting photons in the state $|A\rangle$. Likewise, for the $\{|L\rangle, |R\rangle\}$ basis, it can be shown that P_{AB} corresponds to the probability of detecting photons in the state $|L\rangle$ while $P_{AB'}$ corresponds to the probability of detecting

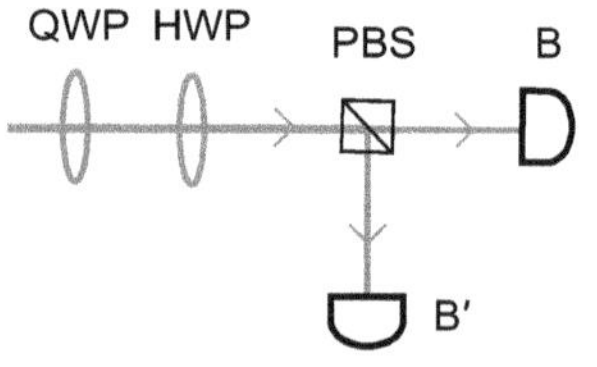

Figure 4.28. A polarizing beam splitter (PBS) and two detectors B and B' are used to perform orthogonal measurements of horizontally and vertically polarized photons. Adding a half-wave plate (HWP) and a quarter-wave plate (QWP), measurement in any basis can be performed by properly orienting the wave plates.

Table 4.2. Commonly used measurement basis for polarization-encoded quantum state estimation.

Measurement basis	Fast axis of QWP	Fast axis of HWP
$\{\|H\rangle, \|V\rangle\}$	0°	0°
$\{\|D\rangle, \|A\rangle\}$	45°	22.5°
$\{\|L\rangle, \|R\rangle\}$	45°	0°

photons in the state $|R\rangle$. We can use these probabilities to estimate the parameters A, B and ϕ of the unknown polarization state.

4.4.3 The experiment

As can be seen by comparing figures 4.29 and 4.25, the layout of this experiment is nearly identical to that of experiment Q2 though few wave plates are added in the signal beam. All the wave plates between the downconversion crystals and the detectors are mounted in motorized rotation stages so that we can use the computer to accurately orient the plates. A photograph is shown in figure 4.30.

The downconversion crystal and the detectors are aligned following the procedures outlined in experiments Q1 and Q2. The laser is vertically polarized and the HWP in the pump beam is also oriented vertically, resulting in downconverted $|HH\rangle$ photons which comprise a separable state. Note that the HWP could also be oriented at an angle of 45° with respect to the vertical, resulting in $|VV\rangle$ downconverted photons but it is a matter of choice.

Using the state generation wave plate (HWP or QWP), different polarization states are generated in the signal beam. The HWP is employed to create linearly polarized states with arbitrary orientation whereas the QWP is used to create elliptically/circularly polarized states. Polarization measurements of identically prepared photons are performed in the three different bases, namely $\{|H\rangle, |V\rangle\}$, $\{|D\rangle, |A\rangle\}$ and $\{|L\rangle, |R\rangle\}$, to determine the probabilities corresponding to equations (4.29), (4.32) and (4.33). The data acquisition time for measurement in each basis is set at two minutes. Parameters of the polarization-encoded quantum state, A, B and ϕ, are determined using equations (4.30)–(4.33). The experimental results show a reasonable agreement with theoretical predictions as summarized in table 4.3. The deviations of measurement from predictions are due to our inability to perfectly prepare the desired states as well as errors in the measurement basis. These imperfect conditions mainly stem from errors in orienting the wave plates in the pump and signal beams. Furthermore, the theoretical predictions rely on the assumption that we are generating pure states, which is not necessarily the case.

This experiment will be revisited in a new light when we discuss experiments Q4 and QST, and where the concept of state purity will be taken up further.

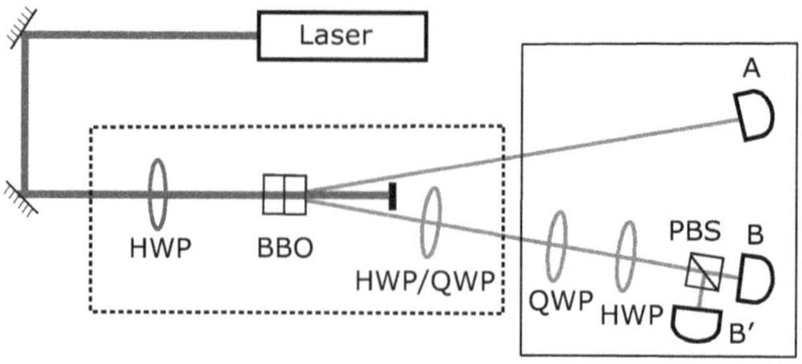

Figure 4.29. Schematic diagram of the quantum state measurement experiment. The dotted box represents the polarization state generator and includes a half-wave plate (HWP), downconversion crystals (BBO) and another wave plate (QWP/HWP). Any arbitrary polarization state can be generated by properly orienting these wave plates. The solid box represents the polarization state analyzer and includes a QWP, an HWP, a polarizing beam splitter (PBS) and three detectors (*A*, *B*, *B*′). Reprinted with permission from [58] © The Optical Society.

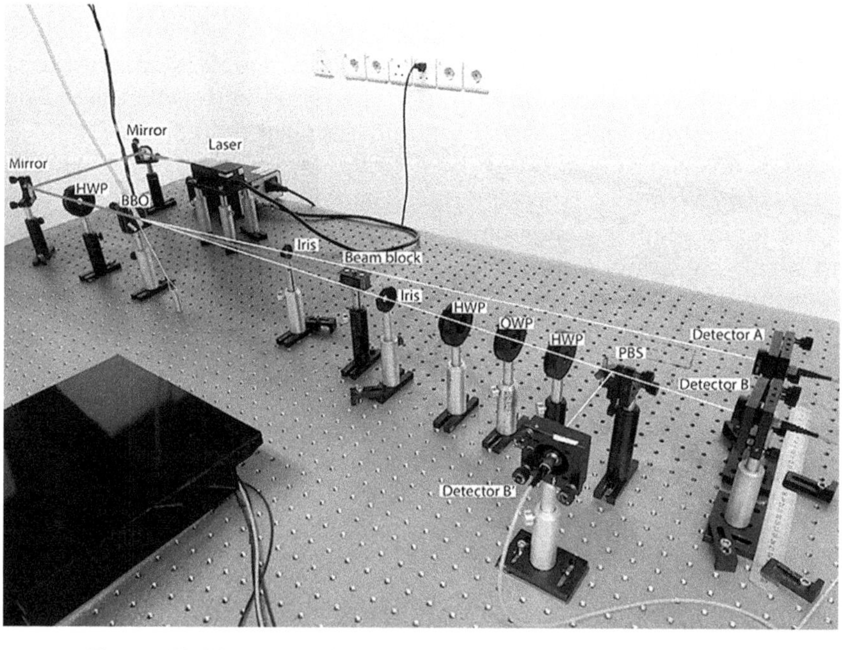

Figure 4.30. Photograph of the quantum state measurement experiment.

4.5 Q4: Visualizing the polarization state of single photons

As discussed in chapter 2, we can use Jones calculus to mathematically express the polarization of light as well as compute the effect of optical elements such as polarizers and wave plates that manipulate the polarization [55, 56]. For single-photon systems, we used a mathematical formalism which is analogous to Jones calculus. We know that Jones calculus is based on the amplitude picture of polarized light, i.e. the electric field. From our discussion so far, we also know that the photodetectors monitor intensity and not the amplitude of light.

Table 4.3. Predicted and measured results of polarization state measurement of single photons. All angles, represented by ϕ, are quoted in radians. The values in parentheses represent uncertainties.

Input	Prediction	Measurement
$\|H\rangle$	$A = 1.000$, $B = 0.000$	$A = 0.993(2)$, $B = 0.120(2)$, $\phi = 1.652(7)$
$\|V\rangle$	$A = 0.000$, $B = 1.000$	$A = 0.254(2)$, $B = 0.967(3)$, $\phi = -1.156(10)$
$\|D\rangle$	$A = 0.707$, $B = 0.707$, $\phi = 0.000$	$A = 0.748(2)$, $B = 0.663(2)$, $\phi = 0.201(3)$
$\|A\rangle$	$A = 0.707$, $B = 0.707$, $\phi = 3.142$	$A = 0.698(1)$, $B = 0.716(1)$, $\phi = 3.055(2)$
$\|L\rangle$	$A = 0.707$, $B = 0.707$, $\phi = 1.571$	$A = 0.732(1)$, $B = 0.681(1)$, $\phi = 1.490(4)$
$\|R\rangle$	$A = 0.707$, $B = 0.707$, $\phi = -1.571$	$A = 0.663(2)$, $B = 0.749(2)$, $\phi = -1.530(5)$

On the other side of the picture, we also briefly discussed the Stokes polarization parameters in chapter 2. Through this formalism, we can characterize the polarization state of light through four intensity parameters [56]. These parameters are often used to characterization the polarization of an optical beam due to the fact that the polarization ellipse [56] is not *directly* measurable [57]. Again, this is because the ellipse is in fact a graphical representation of the *amplitude* formulation of polarized light. Nonetheless, there is a workaround based on the observation that the ellipse parameters are expressible in terms of Stokes parameters and vice versa [56]. See equation (2.19) in chapter 2.

This experiment was originally published in [58]. It is a quantum optical version of the experiment performed in chapter 2 and is motivated from the experiment discussed in [59]. We demonstrate a graphical technique to determine the polarization of single photons, and compare the results with those obtained in experiment Q3. The graphical method is based on well established techniques used in the telecommunications and electrical engineering enterprises that have been used to determine the polarization of antennas. We briefly review this technique in the next section.

4.5.1 Antenna polarimetry and the polarization pattern method

There exists a time-tested empirical technique to determine the polarization of antennas and is known as the 'polarization pattern method' [60]. This method entails the use of a linearly polarized antenna to receive electromagnetic radiation that is being transmitted from another antenna, the polarization profile of which we are interested in. The receiving antenna is positioned normal to the propagation direction of the radiation. The receiving antenna is rotated in a number of steps and the received signal is recorded against the various orientations of the receiving antenna. A peanut-shaped pattern is obtained when one makes a polar plot of the square root of the signal versus the receiving antenna's rotation. This is exemplified in figure 4.31 in which such a pattern circumscribes the polarization ellipse. This peculiar shape is mathematically classified as a hippopede, also known as the 'horse-fetter' [61, 62].

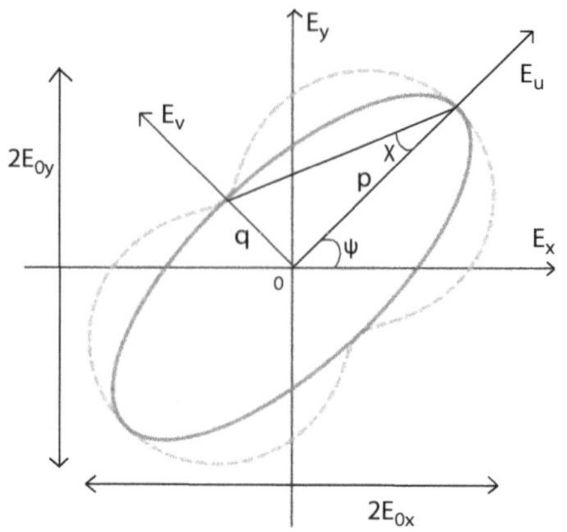

Figure 4.31. Relation of the polarization ellipse and the hippopedal pattern for an arbitrary polarization state. Angular parameters of the ellipse include the orientation angle ψ and the ellipticity angle χ. The ellipticity angle is related to the ratio of the semi-major and semi-minor axes, denoted as p and q respectively. The hippopede is described in the main text. Reprinted with permission from [58] © The Optical Society.

From the figure, it is evident that both the hippopede and the ellipse have the same maximum and minimum radial points the polarization ellipse. Moreover, both closed figures have the same axial ratio (AR) as well as the orientation angle (ψ). The AR is the ratio of semi-major and semi-minor axes. Therefore, by making the corresponding hippopedal plot, one can determine AR and ψ, which are the essential parameters characterizing an arbitrary polarization.

The described method does not fetch us information about the handedness or the helicity of the polarized radiation though. To obtain the helicity information, the receiving antenna needs to be replaced one by one by a left circularly polarized antenna and then a right circularly polarized antenna. The antennas should be of equal gain and the received signal is recorded for both receiver settings. Comparing the two measured signals helps one determine the helicity of the input polarization. For instance, if the signal recoded by the right handed setting is larger than the left handed setting, it implies that the transmitting antenna has a right-handed polarization [63].

4.5.2 Polarization pattern of single photons

For defining the relevant parameters of polarization, it is useful to revisit the polarization ellipse through figure 4.31. We can express an arbitrary ellipse using two angular parameters [56]. There parameters are namely the orientation angle ψ, which can be described as the tilt of the ellipse, and the ellipticity angle χ, which can be determined from the axial ratio of the ellipse. Another angular parameter is the auxiliary angle, which is defined as $\gamma = \tan^{-1}(E_{0y}/E_{0x})$, where E_{0x} and E_{0y} respectively denote the amplitudes of the x and y components of the polarization. This parameter is used in an alternative angular description of the ellipse.

The method of polarization pattern for antennas which we elaborated in the previous section can be readily adapted to an optical system. In the antenna system, a transmitting antenna was responsible for the polarized radiation. For our optical version of the experiment, we can use a scheme similar to that of experiment Q3 for generating an optical beam of arbitrary polarization. Furthermore, instead of using an antenna as an analyzer, we utilize an optical polarization analyzer. Then it is also straightforward to develop a correspondence between linearly polarized antenna and an optical analyzer consisting of a linear polarizer and a photodetector. Finally, the optical replacement of circularly polarized receiving antenna can be a QWP with its axis oriented at ±45°, followed by a horizontally oriented linear polarizer and a photodetector.

Let's mathematically look at this analogy. Take an optical beam of arbitrarily polarized light which is prepared by a state generator (figure 4.32). If we consider the beam to consist of single photons, just as in experiment Q3, we can say that the polarization state encodes the quantum state of single photons, represented by

$$|\Psi\rangle = \cos\gamma|H\rangle + \sin\gamma\, e^{i\phi}|V\rangle. \tag{4.39}$$

Here the angular parameters γ and ϕ suffice to give a complete characterization of the quantum state.

We know that in quantum systems, the analyzer measurements can be computed using Born's rule [3], outlined in chapter 3. A beam of single photons identically prepared in the state $|\Psi\rangle$ is subjected to a linear polarizer oriented at α (labeled P in figure 4.32). The probability of photodetection is then given by

$$\begin{aligned} P(\alpha||\Psi\rangle) &= |\langle\alpha|\Psi\rangle|^2 \\ &= |\,(\cos\alpha\langle H| + \sin\alpha\langle V|)(\cos\gamma|H\rangle + \sin\gamma e^{i\phi}|V\rangle)\,|^2 \\ &= \frac{1}{2}(1 + \cos(2\gamma)\cos(2\alpha) + \sin(2\gamma)\cos(\phi)\sin(2\alpha)), \end{aligned} \tag{4.40}$$

which clearly resembles a Fourier series in α [64] expressed a

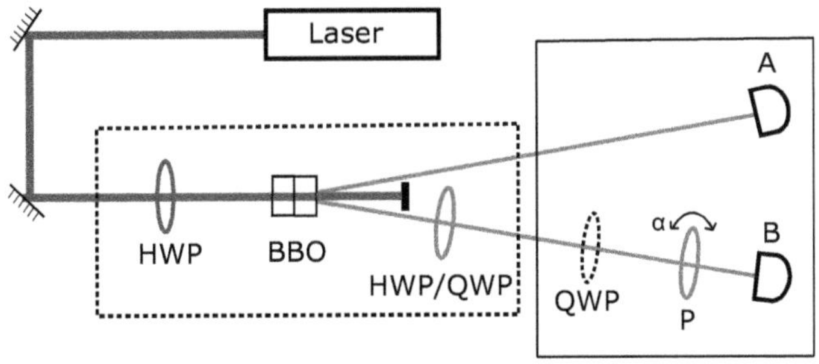

Figure 4.32. Schematic diagram of the quantum state measurement experiment. The dotted box represents polarization state generator and includes a half-wave plate (HWP), downconversion crystals (BBO) and another wave plate (QWP/HWP). Any arbitrary polarization state can be generated by properly orienting these wave plates. The solid box represents polarization state analyzer and includes a polarizer (P) and two detectors (*A* and *B*). An optional QWP can be inserted in the analyzer to obtain the handedness of polarization. Adapted with permission from [58] © The Optical Society.

$$I(\alpha) = \frac{1}{2}(S_0 + S_3\cos(2\alpha) + S_1\sin(2\alpha)), \tag{4.41}$$

where S_0, S_3 and S_1 are the respective Stokes polarization parameters. Although equation (4.40) is for single photons while equation (4.41) is originally described for coherent light, it can be shown that the coefficients in the former equation are indeed expressions for the respective Stokes parameters [56, 64]. Similarly, equation (4.40) can be expressed in terms of the alternative parameters of polarization [65] in the following manner:

$$P(\alpha||\Psi\rangle) = \frac{1}{2}(1 + \cos(2\chi)\cos(2\psi)\cos(2\alpha) + \cos(2\chi)\sin(2\psi)\sin(2\alpha)). \tag{4.42}$$

Comparing the terms of this equation with equation (2.19), we can express equation (4.42) in terms of the Stokes parameters S_0, S_3 and S_1 as

$$P(\alpha||\Psi\rangle) = \frac{1}{2}(S_0 + S_3\cos(2\alpha) + S_1\sin(2\alpha)). \tag{4.43}$$

These Stokes parameters characterize the unknown polarization state of the light leaving the state generator. Therefore, if we have the probability measurements at a number of analyzer orientations, we can make a Fourier curve fit [64, 66] to find the three Stokes parameters S_0, S_3 and S_1.

Now, let's mathematically revisit how the aforementioned polarization pattern method bypasses the need for Fourier curve fitting. From equation (2.19), we know that knowledge of the two angles (χ, ψ) associated with the polarization ellipse lets us immediately calculate the Stokes parameters (S_3 and S_1). Moreover, if we know the sign of χ, we can also uniquely determine the value of S_2 [64].

We get a hippopedal pattern as shown in figure 4.31 if we make a polar plot of $\sqrt{P(\alpha||\Psi\rangle)}$ versus the analyzer angle α. This is essentially a plot between $\sqrt{P(\alpha||\Psi\rangle)}\cos\alpha$ and $\sqrt{P(\alpha||\Psi\rangle)}\sin\alpha$. It is quite evident from figure 4.31 that we can find the angular parameters of the ellipse by plotting just the hippopede based on square root of the α-modulated probability profile. We can directly measure the orientation angle ψ from the hippopedal figure and can calculate the ellipticity angle χ using the AR of the hippopede given by

$$\chi = \pm\tan^{-1}\left(\frac{q}{p}\right), \tag{4.44}$$

where q and p denote the lengths of the semi-minor and semi-major axes of the hippopede. Subsequently, these two angles (χ, ψ) specify the Stokes parameters (S_3 and S_1). However, there still remains ambiguity about the handedness of the polarization. This actually corresponds to the sign of S_2.

Now we need to perform the optical version of the circularly polarized antenna method. We conveniently place a QWP before the polarizer in order to find the handedness of the polarization (figure 4.32). We orient the analyzer QWP at $\pm 45°$

and the polarizer at 0° with respect to the horizontal. The probability expression then reduces to

$$P(L/R||\Psi\rangle) = \frac{1}{2}(1 \mp \sin(2\chi)). \tag{4.45}$$

This test of helicity is based on two projective measurements. If the probability of detecting $|L\rangle$ photons is larger than the probability of detecting $|R\rangle$ photons $(P(L||\Psi\rangle) > P(R||\Psi\rangle))$, the polarization is left handed. In the contrary situation $(P(L||\Psi\rangle) < P(R||\Psi\rangle))$, the polarization is right-handed. If the unknown polarization is linearly polarized, we get $P(L||\Psi\rangle) = P(R||\Psi\rangle)$.

Although we have derived these expressions for a beam of single photons, similar expressions can be derived for coherent light, where we will get intensities instead of probabilities. We can test this classical versus quantum analogy for a single beam of light.

4.5.3 The experiment

The experimental setup, as depicted schematically in figure 4.32 and photographically in figure 4.33, is based on that of experiment Q1, with addition of few optical elements. As in the previous experiments, the detection of the photons striking detector A projects photons of the beam directed at detector B into single-photon states. Using the state generation wave plate (QWP or HWP), we generate a beam of single photons with different polarization profiles and use the polarization pattern method for estimating the generated polarization-encoded quantum states. We

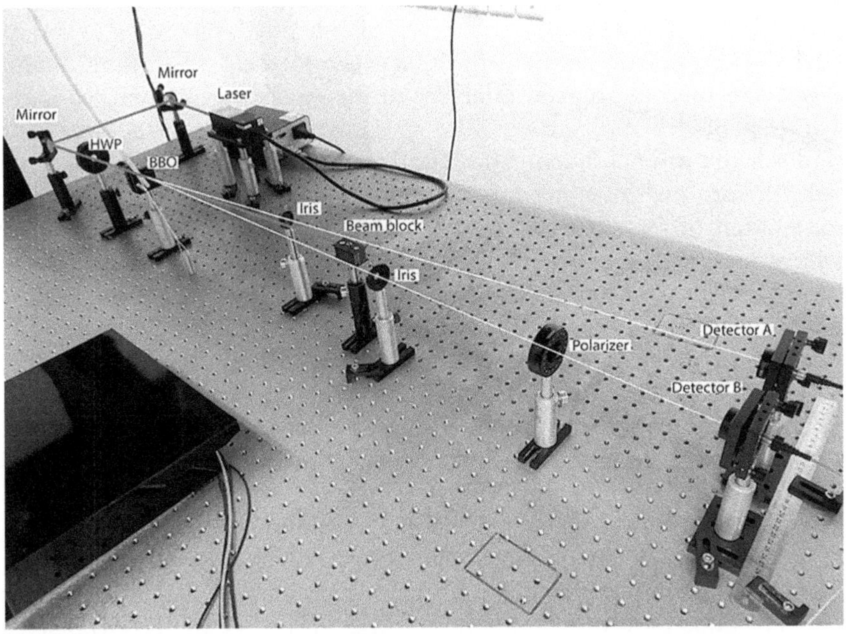

Figure 4.33. Photograph of the polarization pattern measurement setup.

compare the results of the experiment with those of experiment Q3. From the probabilities determined in experiment Q3, we compute the respective Stokes parameters using the following relations [31]:

$$S_0 = P_{|H\rangle} + P_{|V\rangle}, \tag{4.46}$$

$$S_1 = P_{|D\rangle} - P_{|A\rangle}, \tag{4.47}$$

$$S_2 = P_{|L\rangle} - P_{|R\rangle}, \text{ and} \tag{4.48}$$

$$S_3 = P_{|H\rangle} - P_{|V\rangle}. \tag{4.49}$$

Simulated and experimental hippopedal plots for the degenerate polarization states are illustrated in figure 4.34. Making the aforementioned geometric measurements, we determine ellipse parameters from the plots and also compute Stokes parameters for various input polarization states. The results of this quantum state estimation are enlisted in table 4.4. We have performed this analysis for both single counts (beam B considered alone) and coincidence counts (beam B conditioned on beam A). The former corresponds to classical beam of light while the latter corresponds to single-photon states. If we compare the Stokes parameters of the quantum (coincidence counts) and classical (single counts) versions of light, we observe that the polarization properties are similar in both the descriptions of light. This is only true for the case of individual optical beams. It must also be kept in mind that the statistics of photodetection is quite different for the quantum and classical cases, as we discussed in experiment Q2.

The experimental results confirm that we can draw an interesting analogy between optical polarimetry and polarization profiling of antennas. According to this analogy, one can map the optical state generating system to the transmitting antenna and likewise one can map the optical analyzer to the reception antenna. Moreover, the results of this polarimetry experiment agree with those of the quantum state estimation scheme of experiment Q3. Another interesting aspect is the remarkable agreement between the single-photon and coincidence photon results. It is a convincing manifestation of the similarity between classical and quantum regimes of light. The classical picture of light corresponds to coherent light beams such as those of lasers. In our experiment, the beams detected by individual detectors correspond to low intensity classical optical beams. The quantum picture of light of course corresponds to heralded single photons. This experiment and the related discussion can be employed as interesting pedagogical arguments in the laboratory.

A benefit of the technique discussed in this experiment is that from the polar graphs, one can perform few geometric measurements and determine the unknown polarization state. The hippopedal plots are generated by recording photocounts with respect to various orientations of the analyzing polarizer. The geometric measurements fetch us the angular parameters of the ellipse and using these parameters we can compute the three Stokes parameters S_0, S_1 and S_3. To uniquely determine the handedness or helicity of the state, however, we need to place a QWP in the setup and make just two additional measurements.

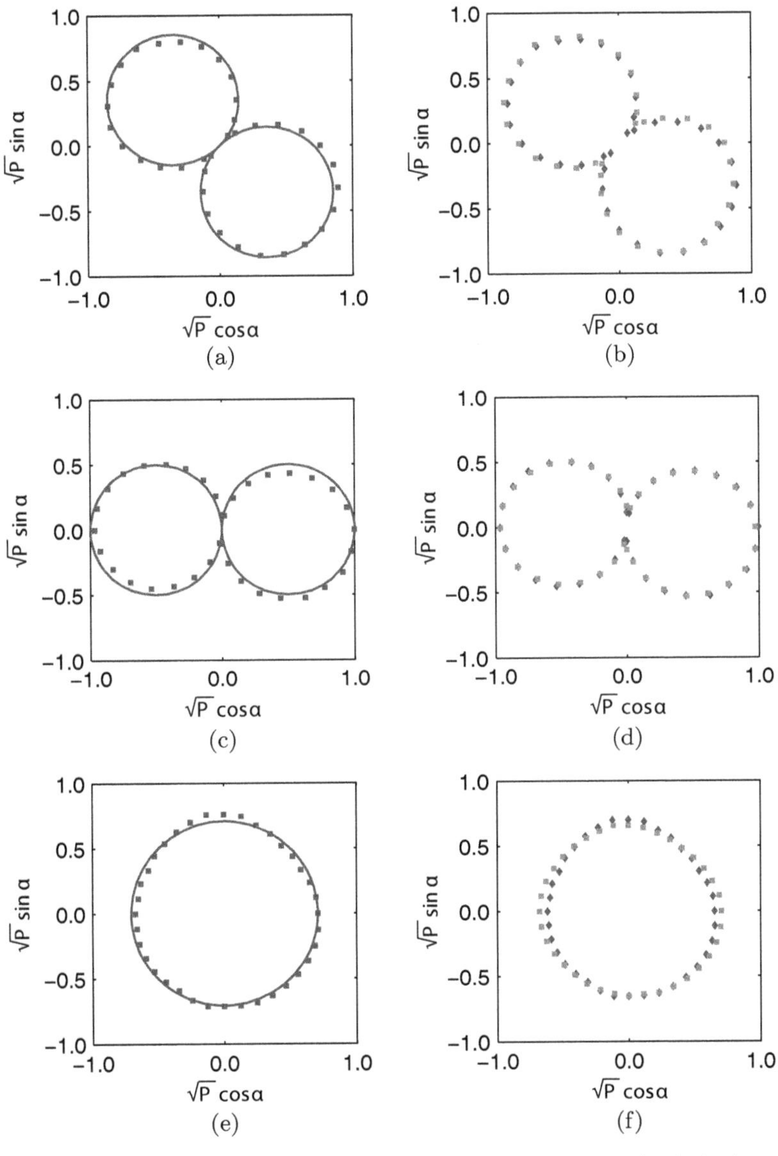

Figure 4.34. Polarization patterns for canonical polarization states. The generated polarization states include anti-diagonal (a), (b), horizontal (c), (d), and left circular (e), (f). (a), (c) and (e) compare simulated plots (smooth curves) with experimental data (discrete points) while (b), (d) and (f) compare polarization state of classical (coherent) light with quantum (single-photon) light. The red points represent measurements for a classical light beam while the blue points represent measurements corresponding to single photons.

4.6 Q5: Single-photon interference and quantum eraser

In experiment Q2, we demonstrated the procedure of creating a beam consisting of single photons. We used detection of the idler beam photons to herald the detection of the signal beam photons. This conditioning on the detections at detector A made

Table 4.4. Experimental results of quantum state estimation and optical polarimetry for canonical polarization states. The column labeled 'Quantum state estimation' quotes the Stokes parameters obtained using the technique discussed in experiment Q3. The columns labeled 'Quantum hippopede' and 'Classical hippopede' enlist the Stokes parameters obtained from the polarization patterns (shown in figure 4.34) using the technique developed in experiment Q4. The quantum hippopede is based on coincidence counts accounting for single photons whereas the classical hippopede is based on individual detector counts accounting for a coherent light beam.

Polarization state	Quantum state estimation	Quantum hippopede	Classical hippopede
$\|H\rangle$	$S_0 = 1.00 \pm 0.00$	1.00 ± 0.04	1.00 ± 0.04
	$S_3 = 0.97 \pm 0.00$	0.95 ± 0.04	0.92 ± 0.04
	$S_1 = -0.06 \pm 0.00$	-0.16 ± 0.04	-0.15 ± 0.04
$\|V\rangle$	$S_0 = 1.00 \pm 0.00$	1.00 ± 0.02	1.00 ± 0.02
	$S_3 = -0.87 \pm 0.00$	-0.98 ± 0.02	-0.92 ± 0.02
	$S_1 = 0.21 \pm 0.00$	-0.08 ± 0.02	-0.09 ± 0.02
$\|D\rangle$	$S_0 = 1.00 \pm 0.00$	1.00 ± 0.00	1.00 ± 0.00
	$S_3 = 0.12 \pm 0.00$	0.04 ± 0.02	0.04 ± 0.02
	$S_1 = 0.69 \pm 0.00$	1.00 ± 0.00	1.00 ± 0.00
$\|A\rangle$	$S_0 = 1.00 \pm 0.02$	1.00 ± 0.02	1.00 ± 0.03
	$S_3 = -0.02 \pm 0.00$	0.11 ± 0.02	0.13 ± 0.03
	$S_1 = -0.90 \pm 0.00$	-0.97 ± 0.02	-0.93 ± 0.03
$\|L\rangle$	$S_0 = 1.00 \pm 0.00$	1.00 ± 0.04	1.00 ± 0.02
	$S_3 = 0.07 \pm 0.00$	-0.04 ± 0.04	0.06 ± 0.02
	$S_1 = 0.08 \pm 0.00$	-0.02 ± 0.04	-0.03 ± 0.02
$\|R\rangle$	$S_0 = 1.00 \pm 0.00$	1.00 ± 0.04	1.00 ± 0.02
	$S_3 = -0.12 \pm 0.00$	-0.07 ± 0.04	0.06 ± 0.02
	$S_1 = 0.03 \pm 0.00$	-0.04 ± 0.06	-0.04 ± 0.02

sure that we obtained single-photon detections at detectors B and B'. This heralded single-photon detection was experimentally established due to a negligible number of coincidence detections between the two detectors that monitored the output beams of a beam splitter. Thus we obtained $g^{(2)}(0) < 1$.

In experiment Q5, we will make these single photons travel through an interferometer and will be able to observe the interference of single photons. In fact, some labs have implemented both second-order correlation measurements (which is our experiment Q2) and interferometric measurements (experiment Q5) simultaneously in the same experiment [3, 35]. However, for the sake of simplicity, we will just focus on the interference part in this experiment.

Before going into the details of the experiment, we give a general overview of interference and quantum erasure and describe some relevant experiments, especially those that can be performed using an SPDC source.

If light is made to pass through an interferometer, the visibility (the extent to which the interference pattern is visible) is dependent on the extent to which the 'which-way' or 'which-path' information is available to the observer or experimenter. We have mentioned this aspect in passing in chapter 1. If, in principle, the experimenter can find out the path taken by the light traversing through the interferometer, no interference will be observed. If the experimenter does not have any idea about the path of light (and cannot even in principle find the which-way information), an interference pattern with very high visibility will be observed. In the more general situation, if partial path information is available, partial interference (having a visibility value between 0 and 1) can be observed [67].

Some experiments have been proposed in which the experimenter can switch between complete information and no information about the path of light by making minor changes in the experimental setup [35, 67, 68]. In such experiments, if the which-path information is unavailable, it is said to be 'erased'—the erasure results in interference fringes of high visibility. Such an interferometer is usually termed as a 'quantum eraser'. In a highly counterintuitive twist to reality, the erasure of which-path information can be materialized even *after* the photons have already passed through the interferometer. This is like altering the outcome of an experiment after it has occurred.

There are a number of ways to make a quantum eraser. For instance, as we discussed in chapter 2, we can insert polarizers into the two paths of a Mach–Zehnder interferometer tagging the paths of the corresponding beams through polarization. The beam exiting the interferometer can be subsequently subjected to a suitable polarization transformation. The polarizers can be oriented to either preserve or erase the which-way information, and thus change the interference visibility.

Interferometry and quantum erasure experiments can also be performed with light sources that produce correlated photon pairs [30, 69]. As we discussed in previous experiments, usually the process of SPDC is employed to generate polarization-entangled photons, conventionally labeled the signal and the idler. In such an experiment, the which-path information can be obtained not only from the signal beam passing through the interferometer but also from the idler beam by performing a suitable measurement. To illustrate, let us assume that the signal and idler photons are correlated in terms of their polarization states. If the photon path through the interferometer is dependent on the photon polarization, a suitable measurement of the idler photon polarization determines the which-path information of the signal photon. Hence, interference fringes are not observed in this case. To observe interference, it must be ensured that the which-path information is erased from both the photon beams. This is realized by modifying the idler polarization measurement.

A number of quantum erasure experiments with setups different from the abovementioned ones have also been presented [70–72]. Following the experiments proposed in [3, 29, 30, 73], we develop an experiment which uses correlated photon pairs to demonstrate quantum erasure in an interferometer. The signal beam traverses the polarization interferometer and falls on a detector, while the idler beam heralds the detection of single photons, ensuring that we are detecting single photons in the signal beam and hence observing single-photon interference. We are interested in observing interference patterns in the measured *coincidence* counts for the two beams. It must be kept in mind that the which-path information needs to be erased in order to see interference. The which-path information is switched on or switched off, i.e. is erased by modifying the orientation of a polarizer which is part of the interferometer.

Before jumping to the practical aspects of this experiment, we will theoretically discuss our polarization interferometer in the following section.

4.6.1 The polarization interferometer and quantum erasure

Consider figure 4.35(a), which shows an arrangement of two beam displacing polarizers (BDPs) and one HWP. If linearly polarized light is input from the left, its horizontally polarized and vertically polarized components are separated in the first BDP, flipped by the HWP and recombined in the second BDP. Changing the tilt of the second BDP changes the path difference of the two beams. This path difference translates into a phase shift ϕ between the vertical and horizontal polarization components of the input light. The effect of this optical arrangement can be encapsulated in the operator

$$\hat{O}_{\rm PI} = \begin{pmatrix} 1 & 0 \\ 0 & e^{i\phi} \end{pmatrix}. \tag{4.50}$$

To vary the input linear polarization of the first BDP, we add another HWP to the left (figure 4.35(b)). Upon exiting the second BDP, the two beams overlap, yet do not still interfere because of having orthogonal polarizations, which are in principle distinguishable. Therefore, to erase the path information, a linear polarizer is placed to the right of the second BDP. Oriented at 45°, the polarizer projects the two overlapping beams into the $|45°\rangle$ state. We refer to the optical assembly of figure 4.35(b) as the polarization interferometer (PI). Assuming the PI input polarization state to be $|H\rangle$, the state leaving the second BDP can be calculated as

$$\begin{aligned} |\Psi\rangle &= \hat{O}_{\rm PI}\hat{O}_{\rm HWP}|H\rangle \\ &= \begin{pmatrix} 1 & 0 \\ 0 & e^{i\phi} \end{pmatrix}\begin{pmatrix} \cos 2\theta & \sin 2\theta \\ \sin 2\theta & -\cos 2\theta \end{pmatrix}\begin{pmatrix} 1 \\ 0 \end{pmatrix} \\ &= \begin{pmatrix} \cos 2\theta \\ \sin 2\theta e^{i\phi} \end{pmatrix}. \end{aligned} \tag{4.51}$$

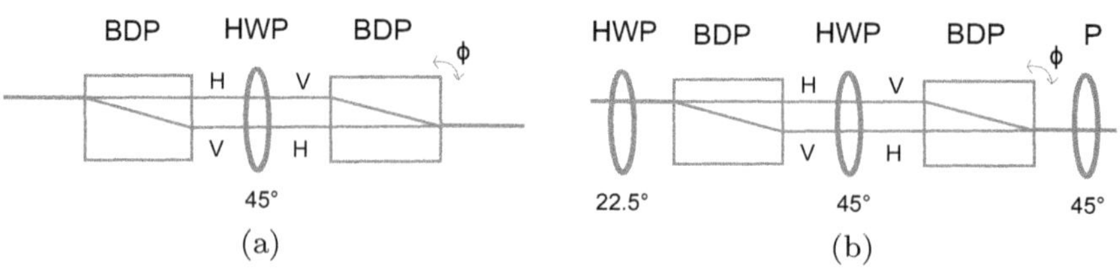

Figure 4.35. Two versions of polarization interferometer. (a) includes two beam displacing polarizers (BDP) and one half-wave plate (HWP). (b) includes an additional HWP and a polarizer (P). $H(V)$ represents horizontally (vertically) polarized light. ϕ represents the phase shift between the H and V components introduced by the interferometer.

The beam with this polarization state passes through the polarizer and falls on a detector. If the first HWP is oriented at $\theta = 22.5°$ and the polarizer is oriented at α, the probability of photodetection is given by

$$P(\alpha) = |\langle\alpha|\Psi\rangle|^2 = \frac{1}{2}(1 + \sin 2\alpha \cos \phi). \tag{4.52}$$

It can be seen that for $\alpha = 45°$, the probability expression reduces to

$$P(22.5°) = \frac{1}{2}(1 + \cos \phi). \tag{4.53}$$

The expression exhibits an oscillatory dependence on ϕ. Hence, an interference pattern is obtained by varying ϕ. In this interferometer setting, the which-way information has been 'erased' and hence interference fringes are predicted. In contrast, for angles such as $\alpha = 0°$ and $\alpha = 90°$, the probability expression yields

$$P(0°) = P(45°) = \frac{1}{2}, \tag{4.54}$$

where the probability of photodetection is independent of the path difference. In other words, changing the path difference will cause no difference in photodetection and there will be no interference fringes. It is because the which-path information of the signal photon becomes available if the photon polarization is known to be either horizontal (in case of $\alpha = 0°$) or vertical (in case of $\alpha = 90°$). So, we have the which-way information and hence no interference is predicted for these cases.

4.6.2 Aligning the interferometer

Let us now set up this interferometer experiment in the single-photon lab. We install the polarization interferometer in the signal beam of the optical setup of experiment Q1.

Make sure that detectors A and B are aligned, as outlined in experiment Q1. The BDPs in the interferometer result in displacement of the beam traveling from the downconversion crystal towards the detector B. Therefore, taking the perspective of detector B to look towards the downconversion source, the detector B is moved 4 mm to the left.

The back-propagation (alignment) laser is shone back through the detector B, as shown in figure 4.36(a), and two irises are aligned close to the downconversion crystal. A BDP is placed in front of detector B. The BDP is positioned and rotated so that two beams emerge parallel from it at a horizontal level. If needed, the detector B is slightly moved sideways so that the BDP beam emerging on the right successfully traverses the two irises and finally falls strikes the center of the downconversion crystal.

An HWP is inserted in front of the BDP and its axis is set at 45°. The second BDP is inserted and rotated ensuring that both the incoming beams recombine and come out as a single output beam. This beam should emerge from the right side of the second BDP and pass back through the irises, as shown in figure 4.36(b).

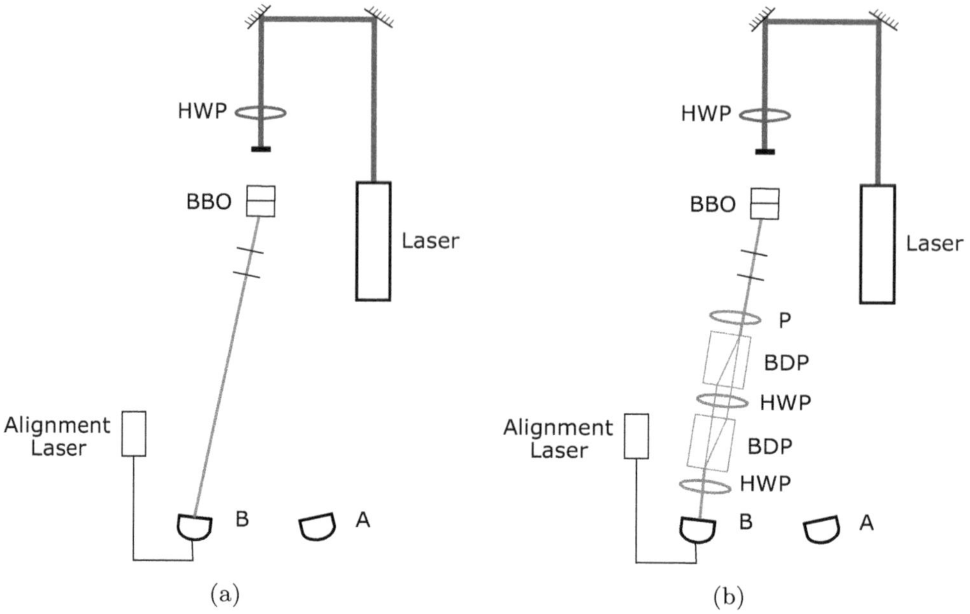

Figure 4.36. Rough alignment of the interferometer. (a) Detector *B* is aligned and two irises are placed near the BBO crystals. (b) Detector *B* is moved 4 mm to the left and the interferometer elements are placed and aligned.

A polarizer oriented at 45° is placed in the combined beam path. A screen can be temporarily inserted in front of the polarizer to see the interference pattern. If is no interference observed, adjust the tilts of both the BDPs. Once interference is visible, slowly adjust the tilts so that there is maximum visibility or contrast of the interference fringes.

The polarizer is replaced with an HWP oriented at 22.5° and the polarizer oriented at 45° is placed before the detector *B*. The alignment laser is removed and the detector *B* is connected with the APD. The rough alignment is complete and now we need to equalize the interferometer path lengths.

Turning the lights off and turning the pump beam and the detectors on, a large range of the BDP tilt is scanned with the stepper motor (connected with the horizontal knob of one of the BDPs). Oscillation of *AB* coincidence counts are observable when the path difference of the two interferometer arms is within the coherence length. When the path lengths are equalized, there will be maximum contrast or visibility of the interference fringes.

If there is no significant change in the *AB* counts while scanning the whole path of the motor, the interferometer should be re-aligned with the alignment laser, as described above, to increase the visibility. Then, scanning the path length difference with the downconverted photons, choose the path in which the visibility is maximum. The visibility may be further improved by very slightly tweaking the tilt of the detector *B* mount or the rotation angle of the wave plates.

4.6.3 The experiment

Our experiment is based on observing the output of the interferometer while varying the orientations of the wave plates. The idea is that for some settings of the interferometric wave plates, the which-path information is available to us. In other words, we know the path that a single photon travels through the interferometer, and in such cases we cannot observe any interference. In contrast, for other settings of the interferometer, the which-path information is unavailable or it is 'erased'. This recovers high-visibility interference fringes.

As mentioned before, this experimental setup is built on the optical setup of experiment Q1. A polarization interferometer is introduced and aligned in the signal beam. Detectors A and B are aligned according to experiment Q1 while the interferometer is aligned according to the previous sections. The schematic of the setup is shown in figure 4.37 while a photograph is shown in figure 4.38. The pump beam HWP is oriented such that the downconverted photons are horizontally polarized. The photons entering the interferometer are horizontally polarized. The HWP between the downconversion crystal and the first BDP is oriented at 22.5° and the HWP between the two BDPs is oriented at 45°.

For different orientations (α) of the polarizer, the coincidence counts N_{AB} are recorded against the path difference controlled by the stepper motor. Two datasets are recorded. The coincidence counts are corrected for accidental coincidences and are plotted as a function of the actuator position (figures 4.39 and 4.40). The visibility of the interference fringes is calculated using the following formula [35] and compared with the theoretical prediction:

$$V = \frac{I_{\max} - I_{\min}}{I_{\max} + I_{\min}} = \frac{N_{\max} - N_{\min}}{N_{\max} + N_{\min}}. \tag{4.55}$$

From figures 4.39 and 4.40, it can be seen that the 45° setting has the highest visibility whereas the 0° and 90° settings have the lowest visibility. Detailed results are summarized in tables 4.5 and 4.6. For the 0° and 90° settings, the path of the photons are known through the interferometer whereas the path information or the which-way information is erased for 45°.

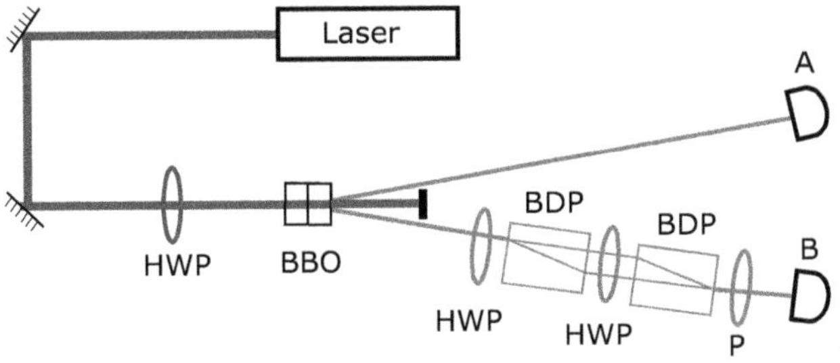

Figure 4.37. Schematic diagram of the single-photon interference and quantum eraser experiment. One of the downconverted photons goes to detector A and heralds the detection of the other photon, which passes through a polarization interferometer and falls on detector B.

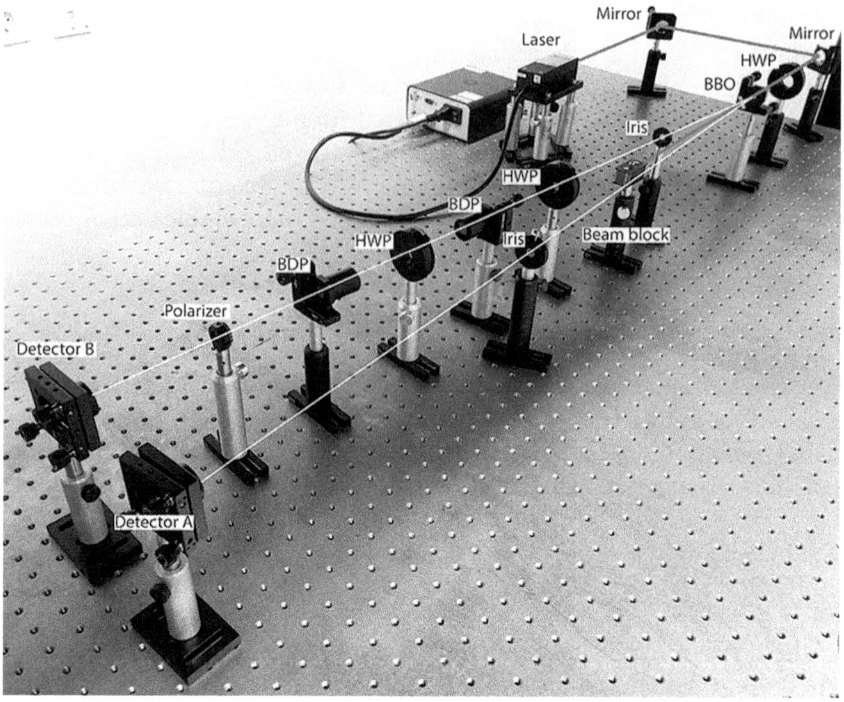

Figure 4.38. Photograph of the single-photon interference and quantum eraser setup.

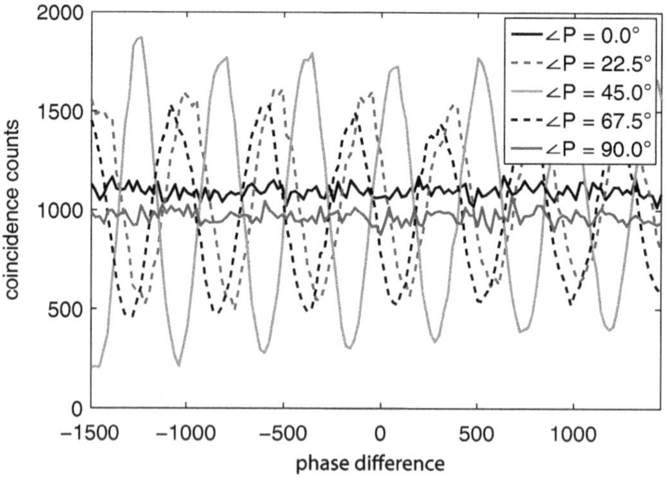

Figure 4.39. Variation of detected coincidence counts with respect to change in phase difference for different polarizer orientations. These plots correspond to the first dataset.

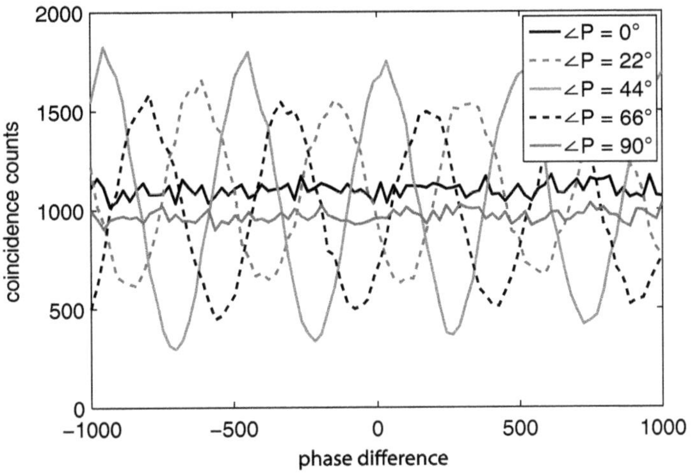

Figure 4.40. Variation of detected coincidence counts with respect to change in phase difference for different polarizer orientations. These plots correspond to the second dataset.

Table 4.5. Visibility of interference for different orientations of the polarizer. These results correspond to the first dataset.

Polarizer orientation (°)	Measured visibility	Predicted visibility
0.0	0.08	0.00
22.5	0.56	0.71
45.0	0.84	1.00
67.5	0.58	0.71
90.0	0.11	0.00

Table 4.6. Visibility of interference for different orientations of the polarizer. These results correspond to the second dataset.

Polarizer orientation (°)	Measured visibility	Predicted visibility
0	0.08	0.00
11	0.30	0.37
22	0.49	0.69
33	0.67	0.91
44	0.78	1.00
55	0.74	0.94
66	0.60	0.74
77	0.39	0.44
90	0.08	0.00

The presence or absence of interference can also be treated in terms of path information, instead of wave interference. Setting the polarizer at 0° effectively gives us knowledge about the path the photon traversed in the interferometer. Therefore the phase difference between the two arms does not have any consequence, and hence no interference pattern is observed. In contrast, if the polarizer is oriented at 45°, there is ambiguity about the which-path information and hence the photon traverses through both paths, which causes its interference with itself.

According to Bohr's principle of complementarity, we cannot simultaneously observe both wave and particle nature of light. Wave-like nature implies an unlocalized behavior whereas particle-like nature implies characteristics of localization. One important subtlety to this principle is that there can be circumstances when one can observe both the behaviors but only partially. If one is able to obtain incomplete or partial information about the path taken by the photon, this results in partial interference. In other words, the interference pattern is there but with compromised visibility. To quote an example, let's say we have the which-path information 40% of the time. Such a situation can arise when our nondestructive measurement is successful only 40% of the time. In this case, the interference pattern will be observed but with a visibility of 60%. The visibility is determined by averaging over a large ensemble of photons. This accounts for the less than ideal visibility of our experimental interference fringes.

The take-home message of this experiment is the bizarre observation that there may be interference even if only a single photon is traversing the interferometer at any instant. This of course may seem very odd because, while doing experiment Q2, we observed the granular nature of light and got used to considering photons 'particles' of light. We do not find particles undergoing interference to be intuitive. One may conclude that although in some ways photons may act like particles of light, we cannot think of them as classical particles in absolute sense. There is much more to them and they can certainly exhibit wave-like behavior too.

We have performed a quantum eraser experiment based on a polarization interferometer. Interference fringes are made to appear by erasing the which-path information and vice versa.

So far, with experiments Q1–Q5, we have explored the quantum state and wave-particle duality associated with one of the two downconverted photons. In chapters 5 and 6, we will investigate other highly non-traditional and often baffling quantum behaviors associated with two polarization-entangled photons.

References

[1] Beck M and Galvez E J 2007 Quantum optics in the undergraduate teaching laboratory *Conf. on Coherence and Quantum Optics* (Washington, DC: Optical Society of America) p CSuA4

[2] Galvez E J and Beck M 2007 Quantum optics experiments with single photons for undergraduate laboratories *Education and Training in Optics* (Washington, DC: Optical Society of America) pp 1–8

[3] Beck M 2012 *Quantum Mechanics: Theory and Experiment* (Oxford: Oxford University Press)

[4] Beck M and Dederick E 2013 Quantum optics laboratories for undergraduates *Education and Training in Optics and Photonics* (Washington, DC: Optical Society of America) p EWB3
[5] Galvez E J and Hamilton N 2004 Undergraduate laboratories using correlated photons: Experiments on the fundamentals of quantum mechanics *Innovative Laboratory Design* pp 113–8
[6] Galvez E J 2014 *Am. J. Phys.* **82** 1018
[7] Galvez E J 2019 Quantum optics laboratories for teaching quantum physics *Education and Training in Optics and Photonics* (Washington, DC: Optical Society of America) p 11143123
[8] Lukishova S G *et al* 2007 Quantum optics teaching laboratory *Conf. on Coherence and Quantum Optics* (Washington, DC: Optical Society of America) p JWC8
[9] Lukishova S G, Stroud C R, Bissell L, Zimmerman B and Knox W H 2008 Teaching experiments on photon quantum mechanics *Frontiers in Optics 2008/Laser Science XXIV/ Plasmonics and Meta-materials/Optical Fabrication and Testing* (Washington, DC: Optical Society of America)
[10] Lukishova S G 2017 Quantum optics and nano-optics teaching laboratory for the undergraduate curriculum: teaching quantum mechanics and nano-physics with photon counting instrumentation *Education and Training in Optics and Photonics* (Washington, DC: Optical Society of America) p 104522I
[11] Branning D, Khanal S, Shin Y H, Clary B and Beck M 2011 *Rev. Sci. Instrum.* **82** 016102
[12] Gea-Banacloche J 2004 Optical realizations of quantum teleportation *Prog. Opt.* **46** 311–54
[13] Kok P *et al* 2007 *Rev. Mod. Phys.* **79** 135
[14] Scarani V *et al* 2009 *Rev. Mod. Phys.* **81** 1301
[15] Aspect A, Dalibard J and Roger G 1982 *Phys. Rev. Lett.* **49** 1804
[16] Gaertner S, Weinfurter H and Kurtsiefer C 2005 *Rev. Sci. Instrum.* **76** 123108
[17] Felekyan S *et al* 2005 *Rev. Sci. Instrum.* **76** 083104
[18] Wahl M, Rahn H-J, Gregor I, Erdmann R and Enderlein J 2007 *Rev. Sci. Instrum.* **78** 033106
[19] Acremann Y, Chembrolu V, Strachan J, Tyliszczak T and Stöhr J 2007 *Rev. Sci. Instrum.* **78** 014702
[20] Dehlinger D and Mitchell M 2002 *Am. J. Phys.* **70** 898
[21] Branning D, Bhandari S and Beck M 2009 *Am. J. Phys.* **77** 667
[22] Masters M F, Heral T and Tummala K 2015 Low-cost coincidence counting apparatus for single photon optics investigations *2015 Conf. on Laboratory Instruction Beyond the First Year of College* (Cambridge, MA: Advanced Laboratory Physics Association) pp 56–9
[23] Branning D and Beck M 2012 An FPGA-based module for multiphoton coincidence counting *Advanced Photon Counting Techniques VI* vol 8375 (Bellingham, WA: International Society for Optics and Photonics) p 83750F
[24] Park B K, Kim Y-S, Kwon O, Han S-W and Moon S 2015 *Appl. Opt.* **54** 4727
[25] Thorn J *et al* 2004 *Am. J. Phys.* **72** 1210
[26] Kwiat P G, Waks E, White A G, Appelbaum I and Eberhard P H 1999 *Phys. Rev.* A **60** R773
[27] Migdall A 1997 *J. Opt. Soc. Am.* B **14** 1093
[28] Grangier P, Roger G and Aspect A 1986 *Europhys. Lett.* **1** 173
[29] Gogo A, Snyder W D and Beck M 2005 *Phys. Rev.* A **71** 052103
[30] Ashby J M, Schwarz P D and Schlosshauer M 2016 *Am. J. Phys.* **84** 95
[31] Altepeter J B, Jeffrey E R and Kwiat P G 2005 *Adv. At. Mol. Opt. Phys.* **52** 105
[32] Brody J and Selton C 2018 *Am. J. Phys.* **86** 412

[33] Dehlinger D and Mitchell M 2002 *Am. J. Phys.* **70** 903
[34] Carlson J, Olmstead M and Beck M 2006 *Am. J. Phys.* **74** 180
[35] Galvez E J *et al* 2005 *Am. J. Phys.* **73** 127
[36] Pearson B J and Jackson D P 2010 *Am. J. Phys.* **78** 471
[37] Stanley R Q 1996 *Am. J. Phys.* **64** 839
[38] Milonni P 1997 *Am. J. Phys.* **65** 11
[39] Burnham D C and Weinberg D L 1970 *Phys. Rev. Lett.* **25** 84
[40] Clauser J F 1974 *Phys. Rev.* D **9** 853
[41] Kimble H J, Dagenais M and Mandel L 1977 *Phys. Rev. Lett.* **39** 691
[42] Funk A and Beck M 1997 *Am. J. Phys.* **65** 492
[43] Brown R H *et al* 1956 *Nature* **177** 27
[44] Twiss R, Little A and Brown R H 1957 *Nature* **180** 324
[45] Glauber R J 1963 *Phys. Rev. Lett.* **10** 84
[46] Glauber R J 1963 *Phys. Rev.* **130** 2529
[47] Glauber R J 1963 *Phys. Rev.* **131** 2766
[48] Sudarshan E 1963 *Phys. Rev. Lett.* **10** 277
[49] Kelley P and Kleiner W 1964 *Phys. Rev.* **136** A316
[50] Mandel L and Wolf E 1965 *Rev. Mod. Phys.* **37** 231
[51] Arecchi F, Gatti E and Sona A 1966 *Phys. Lett.* **20** 27
[52] Clauser J F 1972 *Phys. Rev.* A **6** 49
[53] Greenstein G and Zajonc A 2006 *The Quantum Challenge: Modern Research on the Foundations of Quantum Mechanics* (Burlington, MA: Jones & Bartlett Learning)
[54] Fox M 2006 *Quantum Optics: An Introduction* (Oxford: Oxford University Press)
[55] Jones R C 1941 *J. Opt. Soc. Am.* **31** 488
[56] Collett E 2005 *Field Guide to Polarization* (Bellingham, WA: SPIE)
[57] Wolf E 1954 *Il Nuovo Cimento (1943-1954)* **12** 884
[58] Waseem M H *et al* 2019 *Appl. Opt.* **58** 8442
[59] Mayes T W 1976 *Am. J. Phys.* **44** 1101
[60] Stutzman W L and Thiele G A 1998 *Antenna Theory and Design* (New York: Wiley)
[61] Lawrence J D 2013 *A Catalog of Special Plane Curves* (New York: Dover)
[62] Moroni L 2017 arXiv:1708.00803
[63] Bohnert J 1951 Techniques for handling elliptically polarized waves with special reference to antennas: Part IV-Measurements on elliptically polarized antennas *Proc. IRE* **39** 549–52
[64] Goldstein D H 2016 *Polarized Light* (Boca Raton, FL: CRC Press)
[65] Collett E and Schaefer B 2008 *Appl. Opt.* **47** 4009
[66] Berry H G, Gabrielse G and Livingston A 1977 *Appl. Opt.* **16** 3200
[67] Schwindt P D, Kwiat P G and Englert B-G 1999 *Phys. Rev.* A **60** 4285
[68] Schneider M B and LaPuma I A 2002 *Am. J. Phys.* **70** 266
[69] Walborn S, Cunha M T, Pádua S and Monken C 2002 *Phys. Rev.* A **65** 033818
[70] Herzog T J, Kwiat P G, Weinfurter H and Zeilinger A 1995 *Phys. Rev. Lett.* **75** 3034
[71] Hong C and Noh T 1998 *J. Opt. Soc. Am.* B **15** 1192
[72] Kim Y-H, Yu R, Kulik S P, Shih Y and Scully M O 2000 *Phys. Rev. Lett.* **84** 1
[73] Kwiat P G and Englert B-G 2004 Quantum erasing the nature of reality or, perhaps, the reality of nature Science and Ultimate Reality: Quantum Theory *Cosmology, and Complexity* (Cambridge: Cambridge University Press) p 306

IOP Publishing

Quantum Mechanics in the Single Photon Laboratory

Muhammad Hamza Waseem, Faizan-e-Ilahi and Muhammad Sabieh Anwar

Chapter 5

Experiments related to entanglement and nonlocality

So far, we have discussed the quantum behavior of one-photon systems even though we have two photons, employing one of the beams as a conditioner for the nature of the other beam of photons. From this chapter onward, we will investigate the quantum properties of systems of *two* photons, i.e. the *joint* polarization description of the two-photon system. As our previous practice, we define the following nomenclature to conveniently refer to these two-photon experiments:

- NL1: Freedman's test of local realism.
- NL2: Hardy's test of local realism.
- NL3: CHSH test of local realism.
- QST: Quantum state tomography.

This chapter incorporates experiments NL1, NL2 and NL3 and is about tests of local realism. Certain quantum mechanical systems are predicted to violate local realism as a consequence of entanglement [1], the description of which requires at least two particles. We will recreate some minimalist tests of local realism by employing two beams of photons and two or four detectors. In the following section, we provide an overview of entanglement and local realism. We will discuss experiment QST in chapter 6.

5.1 Entanglement and local realism

Systems that involve two or more particles can demonstrate certain behaviors having no classical counterparts. In particular, the states of two particles, photons in our case, can become 'entangled' with each other. As we discussed briefly in chapter 3, two or more particles are said to be entangled if their combined state cannot be written as a product of single-particle states. In other words, these states are inseparable. Over the last few decades, apart from their central role in

doi:10.1088/978-0-7503-3063-3ch5

discussions of nonlocal quantum correlations [2], entangled particles have been exploited for applications such as quantum cryptography [3, 4], quantum teleportation [5], dense coding [6] and quantum computing [7].

Furthermore, with entangled photons, we can show that quantum mechanics violates local realism. Locality implies that measuring the state of one particle cannot impact the state of another particle. Reality requires that regardless of measurement all measurable quantities in any physical system have definite values [2]. Hence, local realism dictates that if we have a source that is producing pairs of photons, the state of each photon in a pair is definite as soon as it leaves the source. Therefore, any measurement that can be performed on one photon of the pair cannot affect the state of the other member of the pair. Seeing that quantum mechanics does indeed violate local realism as supported by some of the experiments we describe here, we will have to let go of local realism to satisfactorily account for theoretical predictions based on quantum mechanics.

The idea of entanglement was popularized in the physics community through a *gedankenexperiment* (thought experiment) by Einstein, Podolsky and Rosen (EPR), published in 1935 [2]. Entangled photons when individually seen may seem to be randomly polarized. However, the polarization states when seen collectively as a composite bipartite system may show strong correlations, which cannot be described by any classical theory. The Copenhagen interpretation [8] championed by Bohr claims that such correlations arise from the nonlocality associated with measurement, i.e. measuring the state of one particle instantly collapses the entangled state of both particles. The trio EPR in their paper argued that such 'action at a distance' (nonlocality) is impossible and they claimed that quantum mechanics was an incomplete theory because it failed to furnish complete description of reality [2].

However, for many years, this paradox was termed a philosophical debate. Physicists followed the 'shut up and calculate' approach to quantum mechanics and it worked out really well for them. It was John Bell who showed that local realism could in fact be experimentally tested. In 1964, he derived an inequality that must be obeyed by any theory based on local realism [9, 10]. Bell's original inequality was for an idealized system. Later on, other researchers derived similar but testable inequalities [11–13]. These are collectively called Bell's inequalities. Quantum mechanics has violated these inequalities proving that nature is not local realistic [14–16].

In 1972, the first Bell's inequality was tested [14] and this rather elementary inequality was presented by Freedman [11]. The Bell's inequality typically tested in optical systems is the Clauser–Horne–Shimony–Holt (CHSH) inequality [12, 17, 18]. Greenberger, Horne and Zeilinger went a step ahead and showed that an 'all or nothing' test of local realism [19, 20] could be performed. This test was experimentally significantly more difficult and involved three entangled particles instead of two as required in other tests. Eventually, the experiment was performed and its results also agreed with quantum mechanical predictions [15]. In 1993, Hardy derived yet another version of Bell's theorem [13, 21], which is comparatively easier to comprehend as compared to the CHSH test. Experiments based on Hardy's proposal have also been performed and their outcomes too confirm the theoretical predictions of quantum mechanics [16, 22, 23].

In experiments NL1, NL2 and NL3, we will use correlated-photon experiments to study and perform three tests of local realism, based on the ideas of Freedman [11, 14, 24], Hardy [13, 25], and CHSH [12, 17]. Local realism is normally taken for granted because all classical systems are described by it. In contrast, as we are to investigate in this series of experiments, quantum physical systems violate local realistic assumptions. Therefore, we will need to abandon these assumptions in order to account for the outcomes of some experiments.

5.2 The proverbial Alice and Bob experiment

Many analogies have been employed in the literature to describe experiments for testing local realism [26–28]. We will present a description, closely following that of reference [25], which corresponds to our single-photon polarization-based systems.

Suppose we take a source that produces pairs of photons. The two photons of each pair move in different directions: one photon travels to a party named Alice while the other travels to Bob. The source is exactly midway between the two parties, which means that they receive the photons simultaneously. For the sake of simplicity, let us assume that the photons are fired towards the two parties in regular intervals. Each party has a linear polarizer and a detector, as shown in figure 5.1. If Bob aligns his polarizer along a particular angle and his detector picks a photon, he can say that the detected photon was polarized at an angle parallel to his polarizer axis. A similar explanation goes for Alice.

Alice and Bob decide to perform an experiment. Alice randomly orients her polarizer along the two angles θ_{A1} and θ_{A2} and Bob randomly orients his polarizer along θ_{B1} and θ_{B2}. There is a great distance between Alice and Bob and they do not communicate at all while performing the measurements. Once all the measurements have been performed, the two parties meet to check if there are any correlations between their measured photocounts.

Alice and Bob find that the measured photons, when individually seen, seem to be randomly polarized; yet they show unexpectedly strong correlations that cannot be described by any local realist model when seen collectively. This is in fact an optical version of the famous EPR thought experiment. Let us formalize this experiment and then look at it through the lens of Freedman [14], Hardy [13] and CHSH [12].

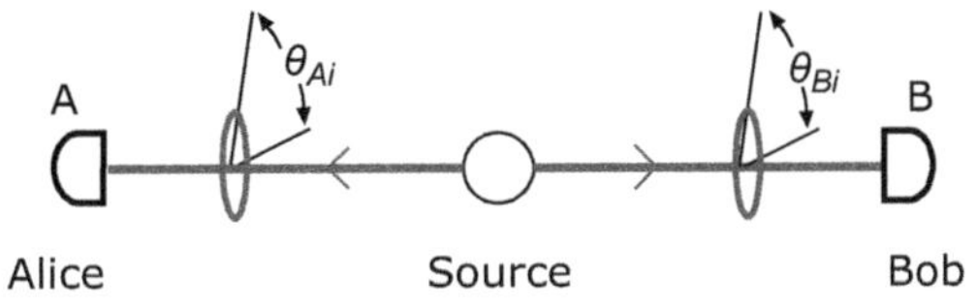

Figure 5.1. The EPR thought experiment in terms of polarization-entangled photons. A source sends out pairs of photons in opposite directions, towards Alice and Bob, who randomly orient their polarizers at angles θ_{Ai} and θ_{Bj} respectively and count the detected photons.

5.3 Generating polarization-entangled photons

Entangled states are an essential ingredient of quantum mechanics and have no classical counterpart [1]. Nowadays, the process of spontaneous parametric down-conversion (SPDC) [23, 29, 30] offers a popular and accessible source of entangled states. This is exactly what we have used in all our experiments.

To understand the polarization-entanglement of photons, let us revisit SPDC. In experiments Q1–Q5, we were not concerned about the polarization state of idler photons, because each idler photon was merely used to herald the detection of the corresponding signal photon. For nonlocality experiments, in contrast, we are required to consider the complete polarization state of the photon pair.

At this point, it is important to re-emphasize that in all our quantum experiments, we are not interested in bunched photons. We use coincidence detection to ensure this *single* photon detection in which each detector click corresponds to the detection of one photon. Now, when we refer to a two-photon system, we mean an experimental system involving two beams of *single* photons.

The polarization-entangled two-photon state is generated using the method proposed by Kwiat [29], in which two type-I downconversion crystals are used such that their optical axes are mutually orthogonal. One of the crystals helps in downconversion of the pump photons that are vertically polarized and the down-converted photon pairs are horizontally polarized. On the other hand, the second crystal is responsible for downconversion of horizontally polarized photons of the pump beam into vertically polarized photon pairs. If we have a pump beam that is polarized at 45° with respect to the horizontal, there is equal likelihood of occurrence of the two downconversion processes (see figure 5.2). On top of that, if we use very thin downconversion crystal, an experimenter monitoring the downconverted photons cannot determine which crystal produced which photon pair. If the photon pairs cannot be distinguished, the two-photon polarization state is an entangled state which is in fact a superposition of the two possible states and is expressed as

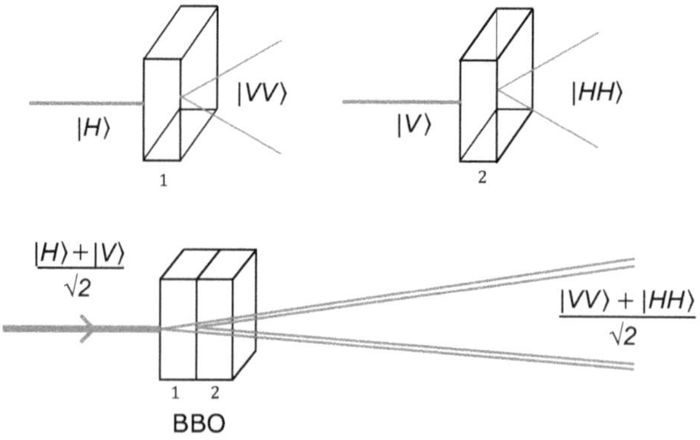

Figure 5.2. Using downconversion to generate entangled states. Crystal 1 downconverts horizontally polarized photons of the pump beam into vertically polarized photon pairs and crystal 2 downconverts vertically polarized photons into horizontally polarized photon pairs. When the two crystals are sandwiched and the input light is in superposition of horizontally and vertically polarized photons, an entangled state is generated.

$$A|H\rangle_A|H\rangle_B + Be^{i\phi}|V\rangle_A|V\rangle_B. \tag{5.1}$$

Here subscript A corresponds to the photon going towards Alice whereas subscript B corresponds to the photon going towards Bob. For the sake of brevity, we can represent this state as

$$A|HH\rangle + Be^{i\phi}|VV\rangle \tag{5.2}$$

and keep in mind that the first member of each ket represent the polarization of Alice's photon whereas the second member represents that of Bob's photon. The desired state with the required A, B and ϕ can be produced by altering the polarization state of the pump beam. This is done by placing a half-wave plate (HWP) and a birefringent quartz plate in the path of the pump beam. The coefficients A and B can be controlled by changing the orientation of the HWP while ϕ can be adjusted by rotating the quartz plate with respect to a vertical axis. Using this arrangement, we can generate states that are arbitrary linear combinations of the states $|HH\rangle$ and $|VV\rangle$.

In experiment Q5, when discussing single-photon interference, we emphasized that to account for the experimental results, we had to consider photons as taking both of the available paths through the interferometer. In a similar fashion, we will find that to explain outcomes of experiments performed using the downconversion source of figure 5.2, we need to interpret the state (5.1) as that of photons being in both states $|HH\rangle$ and $|VV\rangle$ simultaneously. Another way to phrase this is that the photons exist in a superposition of these states. The state (5.1) is pure. In practice, a mixed state is produced which could be an incoherent mixture of the state (5.1) combined with other states. Such a state is described by a density matrix rather than a ket. Experiment QST, which is the subject of chapter 6, allows us to estimate the density matrix.

For an entangled state, the polarization measurements may seem purely random when individually analyzed for the two beams, yet they are perfectly correlated when seen collectively. Neither of the two photons has a well-defined polarization before the measurement. However, if we make a measurement and are able to determine the polarization of either of the photons, the polarization of the other photon is automatically specified. This fact also holds true for measurements performed in other bases [31]. No theory of classical physics can account for such strong correlations in photon polarization. Let's now go into the details of the nonlocality experiments one by one.

5.4 NL1: Freedman's test of local realism

Among the experiments of nonlocality, Freedman's test is the easiest. Although this test is relatively simple, it is historically significant because it was the first experimental test of any Bell's inequality [14]. It was also the focus of Alain Aspect's paper [32] published in 1981. As Freedman's test demands merely three measurements and deriving Freedman's inequality is simpler as compared to other nonlocality tests, it can be used to introduce Bell tests to new students of the subject.

As mentioned before, the standard Bell test performed for optical experiments is the CHSH one [12, 17, 18]. For a two-detector setup, testing the CHSH inequality requires 16 measurements [17] and computing a quantity often denoted as S. Although this serves as a great exercise for an undergraduate laboratory, Freedman's test may be presented earlier to bring home the basic ideas of the Bell tests. Other relatively simple alternatives like the Hardy test usually demand an arrangement of four single-photon detectors [24, 25]. Hardy's test can also be performed with two detectors, but then it requires a larger number of measurements. On the other hand, Freedman's test is performed using only two detectors.

5.4.1 Freedman's inequality

Following the approach in [24], we reproduce the derivation of Freedman's inequality here. This inequality is based on the assumptions of locality, reality and hidden variables, which will be explained along the discussion. Consider six real numbers namely x_1, x_2, y_1, y_2, X, and Y that obey the following relations

$$0 \leqslant x_1 \leqslant X, \tag{5.3}$$

$$0 \leqslant x_2 \leqslant X, \tag{5.4}$$

$$0 \leqslant y_1 \leqslant Y, \tag{5.5}$$

$$0 \leqslant y_2 \leqslant Y, \tag{5.6}$$

$$U \equiv x_1y_1 - x_1y_2 + x_2y_1 + x_2y_2 - Yx_2 - Xy_1. \tag{5.7}$$

Then, the following relation can be shown to hold true [24]:

$$-XY \leqslant U \leqslant 0. \tag{5.8}$$

Now, let's consider photon pairs produced from a source, as shown in figure 5.1. The two photons of any pair take different paths. We further place an analyzer (a polarizer followed by a detector) in each path. Let N_t denote the number of photon pairs directed towards these analyzers in a certain time window. Let $N(\theta_A, \theta_B)$ represent the coincidence counts during the same duration, where θ_A and θ_B denote the orientations of the polarizers, placed in Alice's paths A and Bob's path B respectively. If the measurement time period is sufficiently long, the probability of coincidence detection of the photon pairs is given by

$$P(\theta_A, \theta_B) = \frac{N(\theta_A, \theta_B)}{N_t}, \tag{5.9}$$

i.e. $P(\theta_A, \theta_B)$ represents the ratio of the number of coincidence events and the total number of photon pairs that are being monitored. This probability cannot be directly measured due to nonideal detector efficiencies. We assume that the probability of detecting each photon pair is $P_{12}(\lambda, \theta_A, \theta_B)$, where the subscript refers to photons 1 and 2. We are labeling as 1 and 2 the photons going to the polarizers A

and B respectively. The quantity λ represents a hidden variable, which predetermines how likely a photon pair is to be detected. This photodetection probability may be different for each photon pair. The hidden variable in this case is the photodetection efficiency. Equation (5.9) then describes the average probability of detecting coincidence events for a number of photon pairs. For the sake of argument, let us consider a simple case where $P_{12}(\lambda, \theta_A, \theta_B)$ can either be 0 or 1. This implies that a coincidence event is either detected or not detected and this is predetermined (due to λ) with absolute certainty. In other words, the measurable quantity holds a specific, well-defined value regardless of the question if any entity is actually aware of the value. This is a statement of the reality assumption.

In fact, the probability of coincidence detection $P_{12}(\lambda, \theta_A, \theta_B)$ may have values between 0 and 1, which implies that Freedman's inequality is also applicable to a wider range of hidden variable theories. If $p(\lambda)$ is used to denote the probability distribution of λ, we have

$$P(\theta_A, \theta_B) = \int P_{12}(\lambda, \theta_A, \theta_B) p(\lambda) d\lambda. \tag{5.10}$$

If we assume that polarizer B has no effect on photon 1 and likewise polarizer A has no effect on photon 2, we have

$$P_{12}(\lambda, \theta_A, \theta_B) = P_1(\lambda, \theta_A) P_2(\lambda, \theta_B). \tag{5.11}$$

This is the locality assumption. In addition to locality, this equation also relies on the hidden variable(s) assumption. The common hidden variable λ characterizes both the photons and lets us write the two measurement probabilities independently [24]. In local hidden variable theories, λ captures any correlations in detections of the photon pairs.

We use $P_1(\lambda, \infty)$ to denote the detection probability of photon 1 if no polarizer is placed in its path towards the detector. If we consider the no-enhancement assumption—that adding a polarizer in the path of the light beam cannot increase the number of photodetections—we have

$$0 \leqslant P_1(\lambda, \theta_A) \leqslant P_1(\lambda, \infty). \tag{5.12}$$

Similarly, for a different orientation of polarizer A, we have

$$0 \leqslant P_1(\lambda, \theta_{A'}) \leqslant P_1(\lambda, \infty). \tag{5.13}$$

Likewise, for polarizer B, we have

$$0 \leqslant P_2(\lambda, \theta_B) \leqslant P_2(\lambda, \infty), \tag{5.14}$$

$$0 \leqslant P_2(\lambda, \theta_{B'}) \leqslant P_2(\lambda, \infty). \tag{5.15}$$

Therefore, comparing equations (5.3)–(5.6) with equations (5.12)–(5.15) and using equations (5.7) and (5.8), we can write

$$\begin{aligned} - P_1(\lambda, \infty)P_2(\lambda, \infty) \leqslant P_1(\lambda, \theta_A)P_2(\lambda, \theta_B) - P_1(\lambda, \theta_A)P_2(\lambda, \theta_{B'}) \\ + P_1(\lambda, \theta_{A'})P_2(\lambda, \theta_B) + P_1(\lambda, \theta_{A'})P_2(\lambda, \theta_{B'}) \\ - P_1(\lambda, \theta_{A'})P_2(\lambda, \infty) - P_1(\lambda, \infty)P_2(\lambda, \theta_B) \leqslant 0. \end{aligned} \tag{5.16}$$

After multiplying by $p(\lambda)d\lambda$ and then integrating, equation (5.16) becomes

$$\begin{aligned} - P(\infty, \infty) \leqslant P(\theta_A, \theta_B) - P(\theta_A, \theta_{B'}) + P(\theta_{A'}, \theta_B) \\ + P(\theta_{A'}, \theta_{B'}) - P(\theta_{A'}, \infty) - P(\infty, \theta_B) \leqslant 0. \end{aligned} \tag{5.17}$$

At this point, we make another assumption of rotational invariance, according to which the coincidence counts depend on the relative angle $\phi = |\theta_A - \theta_B|$ between the two polarizers and not on the individual angles, i.e. $P(\theta_A, \theta_B) = P(\phi)$. Hence, we can choose θ_A, $\theta_{A'}$, θ_B, and $\theta_{B'}$ according to the relations expressed as

$$|\theta_A - \theta_B| = |\theta_{A'} - \theta_{B'}| = |\theta_{A'} - \theta_B| = \frac{|\theta_A - \theta_{B'}|}{3} = \phi, \tag{5.18}$$

so that equation (5.17) becomes

$$-P(\infty, \infty) \leqslant 3P(\phi) - P(3\phi) - P(\theta_{A'}, \infty) - P(\infty, \theta_B) \leqslant 0. \tag{5.19}$$

For a polarizer, it is well-known that $P(\phi) = P(\phi + 180°)$. Therefore, taking $P(202.5°) = P(22.5°)$ as well as $\phi = 67.5°$ in equation (5.19), we obtain

$$-P(\infty, \infty) \leqslant 3P(67.5°) - P(22.5°) - P(\theta_{A'}, \infty) - P(\infty, \theta_B) \leqslant 0, \tag{5.20}$$

whereas plugging $\phi = 22.5°$ into equation (5.19) yields

$$-P(\infty, \infty) \leqslant 3P(22.5°) - P(67.5°) - P(\theta_{A'}, \infty) - P(\infty, \theta_B) \leqslant 0. \tag{5.21}$$

Subtracting equation (5.20) from equation (5.21) gives

$$-P(\infty, \infty) \leqslant 4P(22.5°) - 4P(67.5°) \leqslant P(\infty, \infty). \tag{5.22}$$

Multiplying equation (5.22) by N_t, we can rewrite it as

$$-N_0 \leqslant 4N(22.5°) - 4N(67.5°) \leqslant N_0. \tag{5.23}$$

Here $N_0 = N(\infty, \infty)$ represents the number of coincidence measurements if there are no polarizers in the path of the two photon beams and $N(\phi)$ denotes the number of coincidence detections when ϕ is the relative angle between the two polarizers. We can define a parameter δ and rewrite equation (5.23) as

$$\delta = \left| \frac{N(22.5°) - N(67.5°)}{N_0} \right| - \frac{1}{4} \leqslant 0, \tag{5.24}$$

which is a quantity that can be directly determined from the measured photocounts. In other words, if δ is greater than 0, a local realist assumption cannot be correct. This represents Freedman's inequality, which is actually a special case of the Clauser–Horne inequality [33].

This derivation of Freedman's inequality is probably as tedious as a derivation of the CHSH inequality. However, the quantity S in the latter case depends on a set of 16 measurements. Moreover, we cannot express it in a single equation. On the other hand, the quantity δ in Freedman's test is lesser abstract and can be conveniently represented as a single equation.

If all assumptions made while deriving Freedman's inequality were valid, $\delta \leqslant 0$ must hold true. If experimental measurements negate it, at least one of the assumptions is incorrect. The no-enhancement and rotational invariance assumptions can be conveniently tested in the laboratory [24, 33]. The two remaining assumptions are locality and reality. At least one of these must be incorrect if experiments confirm $\delta > 0$.

5.4.2 Quantum mechanical prediction for Freedman's test

The polarization-encoded Bell state we use for Freedman's test can be represented as

$$|\Psi\rangle = \frac{1}{\sqrt{2}}(|H\rangle_A|H\rangle_B + |V\rangle_A|V\rangle_B). \tag{5.25}$$

If our source generates photons in the state $|\Psi\rangle$, the joint probability that Alice detects a photon polarized along angle θ_A and Bob detects a photon polarized along angle θ_B can be determined by Born's rule as

$$P_{\text{ideal}}(\theta_A, \theta_B) = |\langle\theta_A|_A\langle\theta_B|_B|\Psi\rangle|^2 \tag{5.26}$$

if each polarizer is assumed to transmit 100% of the light polarized along its axis. This is a probability of coincidence detection. If the polarizer angles θ_A and θ_B are measured with respect to the horizontal axis of the lab frame, we have

$$|\theta_A\rangle_A = \cos\theta_A|H\rangle_A + \sin\theta_A|V\rangle_A, \tag{5.27}$$

$$|\theta_B\rangle_B = \cos\theta_B|H\rangle_B + \sin\theta_B|V\rangle_B. \tag{5.28}$$

Equation (5.26) then becomes

$$\begin{aligned} P_{\text{ideal}}(\phi) = P_{\text{ideal}}(\theta_A, \theta_B) &= \frac{1}{2}(\cos\theta_A\cos\theta_B + \sin\theta_A\sin\theta_B)^2 \\ &= \frac{1}{2}\cos^2(\theta_A - \theta_B) = \frac{1}{2}\cos^2\phi. \end{aligned} \tag{5.29}$$

If ε_a and ε_b represent the respective transmittances for light polarized along the polarizer axes, the predicted probability of equation (5.29) changes to

$$P_{\text{actual}}(\phi) = \frac{N(\phi)}{N_0} = \frac{1}{2}\varepsilon_a\varepsilon_b\cos^2\phi. \tag{5.30}$$

Hence, we can use equation (5.30) to compute a quantum mechanical prediction for δ as follows [24]:

$$\delta = \left| \frac{N(22.5^\circ) - N(67.5^\circ)}{N_0} \right| - \frac{1}{4} = \frac{\varepsilon_a \varepsilon_b}{2\sqrt{2}} - \frac{1}{4}. \tag{5.31}$$

If we plug $\delta = 0$ in equation (5.31), we get $\varepsilon_a \varepsilon_b = 0.71$ or $\sqrt{\varepsilon_a \varepsilon_b} = 0.84$. This implies that the polarizer transmittances should have a geometric mean of at least 0.84 to demonstrate a violation of Freedman's inequality. Such a kind of requirement on the experimental apparatus is unique to Freedman's test.

5.4.3 Tuning the Bell state

The experimental setup for Freedman's test is shown in figure 5.3 and resembles that of experiment Q1. Therefore, experiment Q1 can be consulted for the alignment procedure. A polarizer is introduced in each path for downconverted photons. An HWP is placed in the pump beam to change the input vertical polarization of light to diagonal, i.e. 45° with respect to the horizontal. For downconversion, two BBO crystals are used. One of the crystal is responsible for downconversion of horizontally polarized photons of the pump beam into pairs of vertically polarized photons, while the other is responsible for downconversion of vertically polarized photons into pairs of horizontally polarized photons. A quartz crystal, which is placed in the pump beam, enables us to effectively minimize the total phase shift between the $|HH\rangle$ and $|VV\rangle$ components of the two-photon state. A mounted quartz plate is shown in figure 5.4(a).

For successful tests of nonlocality, we need to ensure we are working with the required entangled states. For Freedman's test, we would like to create the Bell state given by equation (5.25). Considering the general state (5.1) or (5.2), we would like to get $A = B = 1/\sqrt{2}$ and $\phi = 0$. The former condition dictates that while measuring in the $\{|H\rangle, |V\rangle\}$ basis, we need to obtain $|HH\rangle$ and $|VV\rangle$ photocounts approximately in 1:1 ratio and minimum number of $|HV\rangle$ and $|VH\rangle$ photocounts. The latter condition requires that once we equalize the aforementioned photocounts, on changing measurement basis to $\{|D\rangle, |A\rangle\}$, we should get minimum number of $|DA\rangle$ and $|AD\rangle$ counts, and the $|DD\rangle$ and $|AA\rangle$ counts should also be in the 1:1 ratio. We can check this by calculating the probabilities for the bell state $|\Psi\rangle$ as follows:

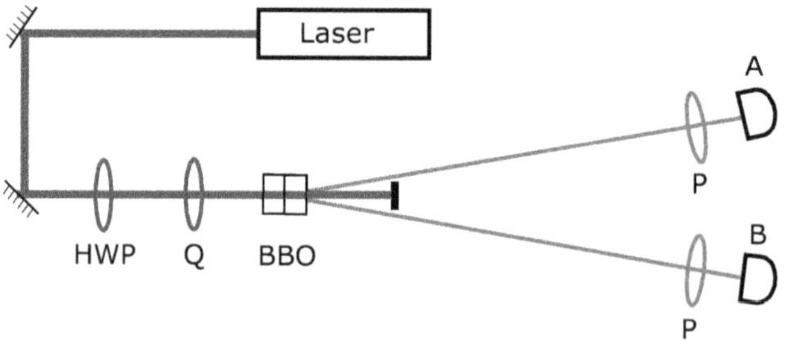

Figure 5.3. Schematic diagram of experimental setup for Freedman's test of local realism. A polarizer (P) is placed in each path of the downconverted photon pair. The pump beam half-wave plate (HWP) and quartz plate (Q) are used to generate Bell state.

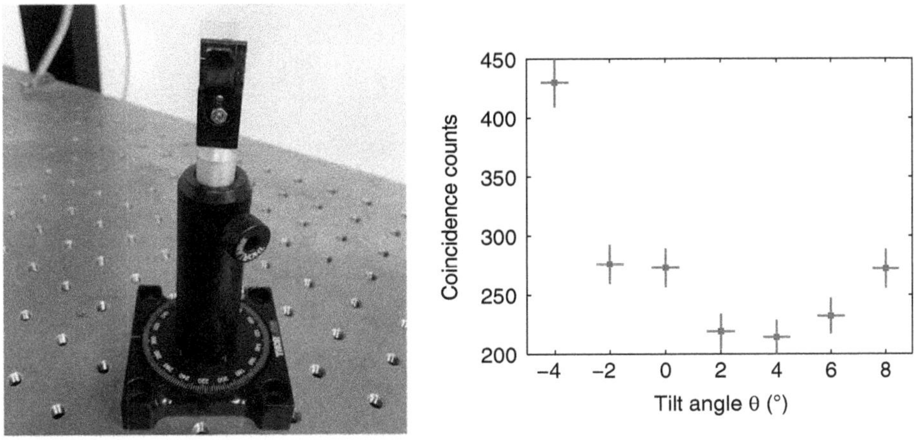

(a) (b)

Figure 5.4. (a) Quartz plate affixed on a mount that is rotatable about the vertical axis. (b) Effect of the tilt of quartz plate on coincidence counts.

$$P(DD) = |(\langle D|_A \langle D|_B)|\Psi\rangle|^2 = \frac{1}{2}, \tag{5.32}$$

$$P(AA) = |(\langle A|_A \langle A|_B)|\Psi\rangle|^2 = \frac{1}{2}, \tag{5.33}$$

$$P(AD) = |(\langle A|_A \langle D|_B)|\Psi\rangle|^2 = 0, \text{ and} \tag{5.34}$$

$$P(DA) = |(\langle D|_A \langle A|_B)|\Psi\rangle|^2 = 0. \tag{5.35}$$

Therefore, in the first step, we equalize the $|HH\rangle$ and $|VV\rangle$ coincidence counts and minimize the $|HV\rangle$ and $|VH\rangle$ counts. Polarizers A and B are set at 0° (to monitor $|HH\rangle$ counts) and the pump beam HWP is adjusted such that the AB counts do not significantly change if both the polarizers are rotated to 90° (to monitor $|VV\rangle$ counts). We also check that for crossed polarizers—one polarizer at 0° and the other at 90°—the AB counts should be minimum. Otherwise, the HWP is adjusted again to meet all these requirements. Essentially the AB counts should be minimum for orthogonal polarizers and comparable for parallel polarizers in both orientations, i.e. at 0° and at 90°.

Next, we change the measurement basis to $\{|D\rangle, |A\rangle\}$ and minimize the phase ϕ. Polarizers A and B are oriented at 45° and −45°. The tilt of the quartz plate Q is adjusted by rotating the plate about the vertical axis until the AB counts are minimized. As shown in figure 5.4(b), these would still not reduce to zero due to experimental imperfections such as accidental counts. After this minimization is achieved, polarizers A and B are now interchanged to −45° and 45° respectively. The AB counts should still be minimum. Otherwise adjust the tilt of Q again.

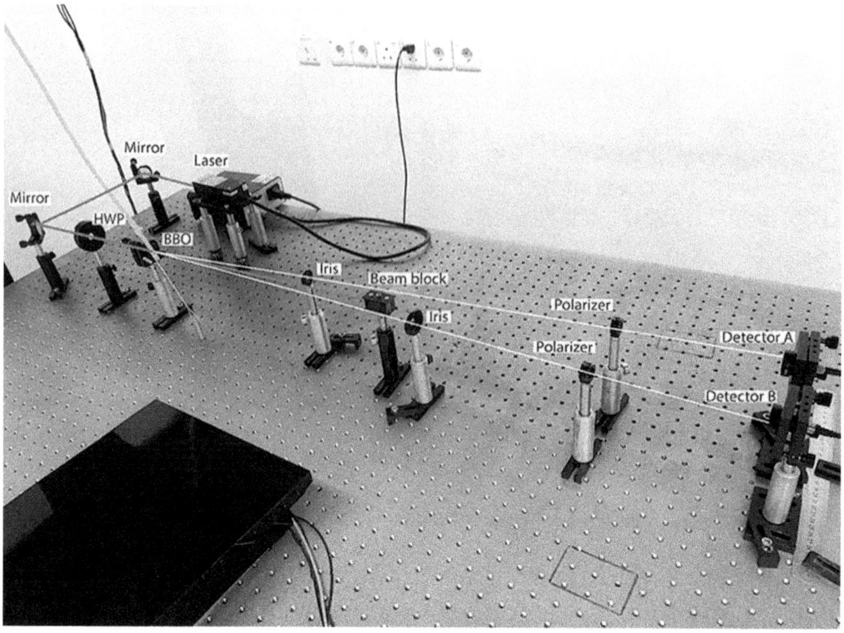

Figure 5.5. Photograph of Freedman's test experiment, excluding the quartz plate.

Very often, we need to iterate between these steps a number of times. Ultimately, AB counts should be equal when the polarizers are parallel and minimized when the polarizers are perpendicular in both $\{|H\rangle, |V\rangle\}$ and $\{|D\rangle, |A\rangle\}$ bases. Once this is done, the state of equation (5.25) is approximately achieved and Freedman's test can be performed.

5.4.4 The experiment

A photograph of the experimental setup is shown in figure 5.5. As explained in the previous section, the following Bell state is created by adjusting the pump beam HWP and the quartz plate:

$$|\psi\rangle = \frac{1}{\sqrt{2}}(|H\rangle_A|H\rangle_B + |V\rangle_A|V\rangle_B). \tag{5.36}$$

As discussed previously and shown in figure 5.4(b), the tilt of the quartz plate is optimized for minimal counts in the crossed polarizer setting.

The coincidence counts $N(22.5°)$, $N(67.5°)$ and N_0 are measured over equally timed intervals. To determine the polarizer transmittances, the coincidence counts N_a and N_b are measured with one polarizer removed at a time and the other oriented for maximum counts. The transmittances are calculated as $\varepsilon_a = 2N_a/N_0$ and $\varepsilon_b = 2N_b/N_0$ [24]. There is a factor of 2 in these expressions because for the Bell state (5.25), placing a linear polarizer in one of the downconverted beams reduces the coincidence counts to roughly half the original value N_0. The reduced counts N_a and N_b are proportional to the respective transmittances.

We record two datasets. The measurements clearly violate Freedman's inequality and the results are summarized in table 5.1. The two values of δ experimentally achieved are greater than 0 and in fact deviate from the local realistic prediction by 18 and 29 standard deviations respectively.

Using the polarizer transmittances $\varepsilon_a = 0.97$ and $\varepsilon_b = 0.92$, and from equation (5.31), the quantum mechanical prediction for δ is calculated as

$$\delta = \frac{\varepsilon_a \varepsilon_b}{2\sqrt{2}} - \frac{1}{4} = \frac{0.97 \times 0.92}{2\sqrt{2}} - \frac{1}{4} = 0.064. \tag{5.37}$$

The theoretical value is lower than the experimental δ as shown in table 5.1. This may be due to nonideal behavior of the polarizers at certain orientations.

We have seen that we need to use just two single-photon detectors and perform only three measurements to perform a test of Freedman's inequality. It is perhaps the most economical test for Bell-type nonlocality. On the other hand, this test has certain limitations as well. For instance, it makes use of the no-enhancement and the rotational invariance assumptions (which are related to properties of the polarizers) while deriving the inequality. A stronger test of nonlocality, such as the CHSH one, will not require these additional assumptions. The additional assumptions are said to create 'loopholes' which make Freedman's test a rather weak test of local realism. Nonetheless, the minimalist hardware requirements of this nonlocality test still make it an excellent experiment for teaching purposes. Furthermore, in a pedagogical context, discussing the limitations of this experiment offers a great opportunity to talk about other subtler loopholes [34–36] which have been closed a few years ago. For example, a *communication loophole* exists if the result of Bob's measurement could in principle be communicated to Alice's setup in time to affect her measurement. It goes without saying that this loophole is present in all our nonlocality experiments. We refer the interested reader to this review article [37] on loopholes in Bell-type nonlocality tests.

5.5 NL2: Hardy's test of local realism

We will observe that implementation of Hardy's test [25] is quite similar to that of the CHSH test of local realism [17, 30]. We will see that the apparatus used in both the experiments is the same, but there are some differences in how these experiments are performed. For instance, different entangled states are generated with the downconverted photons and the analyzing polarizers are oriented at different sets of angles. Most importantly, in Hardy's test, we are interested in analyzing the

Table 5.1. Results of Freedman's test of local realism.

Integration time	$N(22.5°)$	$N(67.5°)$	N_0	δ
20 s	339 ± 4	12 ± 1	936 ± 10	0.099 ± 0.005
120 s	355 ± 2	13 ± 0	991 ± 7	0.095 ± 0.003

coincidence counts to determine a quantity represented as H instead of a different quantity used in the CHSH test.

One may then question the preference of performing one test to the other. Even though it is incrementally more involved than Freedman's test, Hardy's test is considerably easier to comprehend as compared to the CHSH test. Therefore, before attempting the CHSH test, it makes sense to explore Hardy's test in the single-photon quantum mechanics laboratory.

5.5.1 The Hardy inequality

We derive the Hardy inequality following the approach detailed in [31]. For this derivation, let us go back to Alice and Bob (figure 5.1), who are interested in the joint probability $P(\theta_A, \theta_B)$ of the event when Alice detects a photon polarized along the angle θ_A and Bob detects a photon polarized along the angle θ_B.

According to quantum mechanics, the aforementioned probability is encoded in the composite wavefunction $|\Psi\rangle$. On the other hand, in local realist theories, we have already proposed the presence of a hidden variable λ which is associated with the source and which determines the joint probabilities. As mentioned earlier in this chapter, this is called a hidden variable because it cannot be measured, yet it encodes the probabilities we want to determine. According to the hidden variable theory,

$$P(\theta_A, \theta_B) \equiv \sum_{\lambda} P(\theta_A, \theta_B, \lambda) = \sum_{\lambda} P(\theta_A, \theta_B|\lambda)P(\lambda) \tag{5.38}$$

since there can be more than one hidden variable, specifying the joint probability. This is the reality assumption. The probability can also be written as

$$\begin{aligned} P(\theta_A, \theta_B) &= \sum_{\lambda} P(\theta_A, \theta_B, \lambda) = \sum_{\lambda} P(\theta_A|\theta_B, \lambda)P(\theta_B, \lambda) \\ &= \sum_{\lambda} P(\theta_A|\theta_B, \lambda)P(\theta_B|\lambda)P(\lambda). \end{aligned} \tag{5.39}$$

Another assumption here is that the probability distributions we are dealing with are all classical probability distributions, i.e. the probabilities are real, nonnegative and normalized.

Furthermore, the locality assumption tells us that measurement made by Bob cannot affect measurement made by Alice and vice versa. Mathematically, this can be stated as

$$P(\theta_A|\theta_B, \lambda) = P(\theta_A|\lambda) \text{ and} \tag{5.40}$$

$$P(\theta_B|\theta_A, \lambda) = P(\theta_B|\lambda). \tag{5.41}$$

Therefore, following local realism, the joint probability of equation (5.39) becomes

$$P(\theta_A, \theta_B) = \sum_{\lambda} P(\theta_A|\lambda)P(\theta_B|\lambda)P(\lambda). \tag{5.42}$$

The normalization of probability distributions tells us that

$$P(\theta_A|\lambda) + P(\theta_A^{\perp}|\lambda) = 1, \tag{5.43}$$

meaning that if the source produces photons specified by the hidden variable λ, two possibilities exist for Alice's photon: the photon is polarized either along θ_A or along $\theta_A^{\perp}$ (i.e. perpendicular to θ_A). The original joint probability can then be treated as follows:

$$\begin{aligned}
P(\theta_{A1}, \theta_{B1}) &= \sum_{\lambda} P(\theta_{A1}|\lambda)P(\theta_{B1}|\lambda)P(\lambda) \\
&= \sum_{\lambda} P(\theta_{A1}|\lambda)P(\theta_{B1}|\lambda)(P(\theta_{A2}|\lambda) + P(\theta_{A2}^{\perp}|\lambda))P(\lambda) \\
&= \sum_{\lambda} P(\theta_{A1}|\lambda)P(\theta_{B1}|\lambda)P(\theta_{A2}|\lambda)P(\lambda) \\
&\quad + \sum_{\lambda} P(\theta_{A1}|\lambda)P(\theta_{B1}|\lambda)P(\theta_{A2}^{\perp}|\lambda)P(\lambda) \\
&= \sum_{\lambda} P(\theta_{A1}|\lambda)P(\theta_{B1}|\lambda)P(\theta_{A2}|\lambda)(P(\theta_{B2}|\lambda) + P(\theta_{B2}^{\perp}|\lambda))P(\lambda) \\
&\quad + \sum_{\lambda} P(\theta_{A1}|\lambda)P(\theta_{B1}|\lambda)P(\theta_{A2}^{\perp}|\lambda)P(\lambda) \\
&= \sum_{\lambda} P(\theta_{A1}|\lambda)P(\theta_{B1}|\lambda)P(\theta_{A2}|\lambda)P(\theta_{B2}|\lambda)P(\lambda) \\
&\quad + \sum_{\lambda} P(\theta_{A1}|\lambda)P(\theta_{B1}|\lambda)P(\theta_{A2}|\lambda)P(\theta_{B2}^{\perp}|\lambda)P(\lambda) \\
&\quad + \sum_{\lambda} P(\theta_{A1}|\lambda)P(\theta_{B1}|\lambda)P(\theta_{A2}^{\perp}|\lambda)P(\lambda).
\end{aligned} \tag{5.44}$$

If we keep in mind the basic axioms that all probabilities need to be real, positive and less than 1, it is not hard to see that the three terms in equation (5.44) obey the inequalities

$$\begin{aligned}
\sum_{\lambda} P(\theta_{A1}|\lambda)P(\theta_{B1}|\lambda)P(\theta_{A2}|\lambda)P(\theta_{B2}|\lambda)P(\lambda) &\leqslant \sum_{\lambda} P(\theta_{A2}|\lambda)P(\theta_{B2}|\lambda)P(\lambda) \\
&= P(\theta_{A2}, \theta_{B2}),
\end{aligned} \tag{5.45}$$

$$\begin{aligned}
\sum_{\lambda} P(\theta_{A1}|\lambda)P(\theta_{B1}|\lambda)P(\theta_{A2}|\lambda)P(\theta_{B2}^{\perp}|\lambda)P(\lambda) &\leqslant \sum_{\lambda} P(\theta_{A2}|\lambda)P(\theta_{B2}^{\perp}|\lambda)P(\lambda) \\
&= P(\theta_{A2}, \theta_{B2}^{\perp}),
\end{aligned} \tag{5.46}$$

$$\begin{aligned}
\sum_{\lambda} P(\theta_{A1}|\lambda)P(\theta_{B1}|\lambda)P(\theta_{A2}^{\perp}|\lambda)P(\lambda) &\leqslant \sum_{\lambda} P(\theta_{B1}|\lambda)P(\theta_{A2}^{\perp}|\lambda)P(\lambda) \\
&= P(\theta_{A2}^{\perp}, \theta_{B1}).
\end{aligned} \tag{5.47}$$

Putting these inequalities into equation (5.44) yields the inequality

$$P(\theta_{A1}, \theta_{B1}) \leqslant P(\theta_{A2}, \theta_{B2}) + P(\theta_{A1}, \theta_{B2}^{\perp}) + P(\theta_{A2}^{\perp}, \theta_{B1}). \tag{5.48}$$

This is called the Bell–Clauser–Horne inequality and it should be obeyed by any local realistic theory. If this inequality is violated, quantum mechanical predictions jeopardizing a local realistic theory are validated and nature is shown to violate local realism at the fundamental level.

The above inequality can be also written as

$$P(\theta_{A2}, \theta_{B2}) \geqslant P(\theta_{A1}, \theta_{B1}) - P(\theta_{A1}, \theta_{B2}^{\perp}) - P(\theta_{A2}^{\perp}, \theta_{B1}). \tag{5.49}$$

This involves four independent angles ($\theta_{A1}, \theta_{A2}, \theta_{B1}, \theta_{B2}$). If we choose these angles to be $\theta_{A1} = \beta$, $\theta_{B1} = -\beta$, $\theta_{A2} = -\alpha$ and $\theta_{B2} = \alpha$, equation (5.49) becomes

$$P(-\alpha, \alpha) \geqslant P(\beta, -\beta) - P(\beta, \alpha^{\perp}) - P(-\alpha^{\perp}, -\beta). \tag{5.50}$$

If we define H as

$$H = P(\beta, -\beta) - P(\beta, \alpha^{\perp}) - P(-\alpha^{\perp}, -\beta) - P(-\alpha, \alpha), \tag{5.51}$$

then, if we experimentally obtain $H \leqslant 0$, local realism is satisfied. In contrast, if we obtain $H > 0$, local realism is violated. Let us see how a strictly quantum mechanical theory allows H to be positive and hence violate local realism.

5.5.2 Quantum mechanical prediction for Hardy's test

Let us consider a bipartite system defined by the general state

$$|\Psi\rangle = A|H\rangle_A|H\rangle_B + B|V\rangle_A|V\rangle_B, \tag{5.52}$$

with A and B being real numbers and the state being normalized (i.e. $|A|^2 + |B|^2 = 1$).

If the source generates photons in the state $|\Psi\rangle$, the joint probability of the event that Alice detects a photon polarized along the angle θ_{Ai} and Bob detects a photon polarized along the angle θ_{Bj} is given by Born's rule as

$$P(\theta_{Ai}, \theta_{Bj}) = |(\langle\theta_{Ai}|_A\langle\theta_{Bj}|_B)|\Psi\rangle|^2. \tag{5.53}$$

The measurement basis is chosen by orienting a polarizer at the appropriate angle in the path of each beam. If Alice's and Bob's polarizers are oriented at θ_{Ai} and θ_{Bj} with respect to the horizontal, we can use the formulation

$$|\theta\rangle = \cos\theta|H\rangle + \sin\theta|V\rangle \tag{5.54}$$

to determine the joint probability as

$$\begin{aligned} P(\theta_{Ai}, \theta_{Bj}) = &|((\langle H|_A \cos\theta_{Ai} + \langle V|_A \sin\theta_{Ai}) \\ &(\langle H|_B \cos\theta_{Bj} + \langle V|_B \sin\theta_{Bj}) \\ &(A|H\rangle_A|H\rangle_B + B|V\rangle_A|V\rangle_B)|^2 \\ = &(A\cos\theta_{Ai}\cos\theta_{Bj} + B\sin\theta_{Ai}\sin\theta_{Bj})^2. \end{aligned} \tag{5.55}$$

Using equation (5.55), one can verify that maximal violation of Hardy's inequality can be achieved if either of the following two states are generated to perform the experiment [25, 31]:

$$|\Psi_1\rangle = \sqrt{0.8}\,|H\rangle_A|H\rangle_B + \sqrt{0.2}\,|V\rangle_A|V\rangle_B, \tag{5.56}$$

$$|\Psi_2\rangle = \sqrt{0.2}\,|H\rangle_A|H\rangle_B + \sqrt{0.8}\,|V\rangle_A|V\rangle_B. \tag{5.57}$$

For $|\Psi_1\rangle$, the analysis angles are $\alpha = 55°$ and $\beta = 71°$, and for $|\Psi_2\rangle$, the analysis angles are $\alpha = 35°$ and $\beta = 19°$ [25]. We can check that for $A = \sqrt{0.2}$ and $B = \sqrt{0.8}$ in equation (5.52), choosing $\alpha = 35°$ and $\beta = 19°$, equations (5.51) and (5.55) result in $H = 0.093 > 0$. Hence, mathematical analysis shows that the system will violate local realism. We will now describe how this result can be experimentally verified.

5.5.3 The experimental setup

As far as the detection scheme is concerned, this experiment can be set up in two ways. We can employ either a two-detector or a four-detector scheme. If we use the two-detector scheme, we are required to perform measurements in 16 different settings of the analyzing polarizers. On the other hand, with the four-detector scheme, we need to perform measurements with just 4 settings of the analyzer systems. Hence, to obtain results similar to those of the four-detector setup, the two-detector setup demands a four times longer duration for data acquisition. We use the four-detector scheme as shown in figure 5.6.

The experimental setup is shown in figure 5.6 and can be built on the setup of experiment Q2, following the guidelines of which, the fourth detector A' can also be aligned. Unlike the setup of Freedman's experiment, this is a four-detector setup in which photons of each downconverted beam pass through an HWP and a polarizing beam splitter (PBS). We can change the HWP orientation which is just equivalent to modifying the polarization axis of the PBS. Finally, the photons leaving the PBS are monitored by the respective detectors. Now we discuss alignment of the detectors and the downconversion crystals.

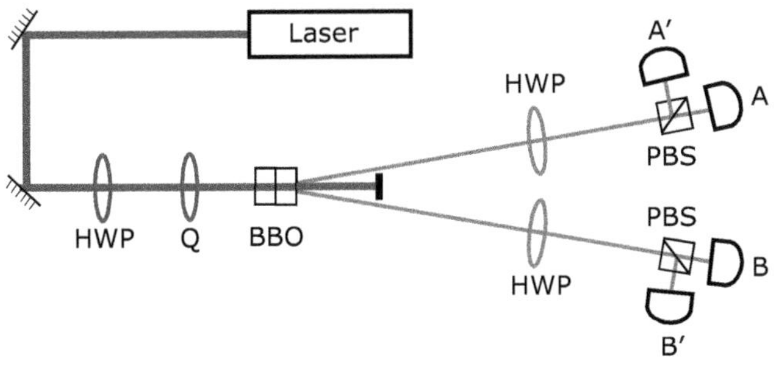

Figure 5.6. Schematic diagram of experimental setup for Hardy's test. In each path of the downconverted photon pair, a half-wave plate (HWP), a polarizing beam splitter (PBS) and two detectors are placed. The pump beam half-wave plate (HWP) and quartz plate (Q) are used to generate the Hardy state.

5.5.4 Aligning the detectors

The ultimate requirement for successfully conducting a nonlocality experiment is tuning the downconversion source so that the required entangled state is generated. For Hardy's test, we aim to generate a state that closely resembles either of the two states expressed in equations (5.56) or (5.57).

We know that for an arbitrary linearly polarized pump beam, our downconversion crystals generate an entangled state (5.2). The pairs of horizontally polarized downconverted photons are then detectable with probability $|A|^2$ and on the other hand pairs of vertically polarized photons can be detected with probability $|B|^2$. Just as we saw in Freedman's test, tuning this state requires tuning the parameters A, B and ϕ. We can adjust the relative magnitudes of A and B by changing the orientation of the pump beam HWP. Subsequently, we can also control the relative phase ϕ by adjusting the tilt the quartz plate with respect to the vertical axis of the lab frame. We desire to tune this phase to be as close as possible to 0° [29]. Before creating this state, we need to align the optical setup, which is detailed as follows.

First, we orient the pump beam HWP to obtain a vertically polarized pump beam. This ensures pumping of just one of the downconversion crystals, and in this case the downconverted photons are horizontally polarized. Following the procedure outlined in experiment Q1, we align the crystal as well as the detectors A and B. Furthermore, revisiting the procedure of experiment Q2, we align the PBS, the detector HWPs, and the detectors A' and B'.

By now, our experimental setup is aligned with respect to one of the downconversion crystals. When we optimize the alignment with respect to one crystal, we need to tweak the tilt of the crystal along only one axis. It is because if we are using a vertically polarized pump beam, the crystal responding to this polarization is only sensitive to tilt adjustments made in the vertical direction. Therefore, for alignment with respect to the second crystal, we change the orientation of the pump beam HWP by 45°. This modification results in a horizontally polarized pump beam which is downconverted by the second crystal. We tweak the tilt of the crystal pair along the horizontal direction to get maximal count rates. It should be kept in mind that only the second crystal is sensitive to this tilt, and therefore the previous alignment with respect to the first crystal will not be significantly affected. Once the horizontal and vertical tilts of the downconversion crystals are optimized, the setup is reasonably aligned for the experiment.

Next, we need to orient the pump HWP and tweak the tilt of the birefringent quartz plate so that we generate the required polarization-entangled state admitting Hardy's test.

5.5.5 Tuning the Hardy state

For our experiment, we would like to tune the entangled state $|\Psi_2\rangle$ given in equation (5.57). Following the method outlined in [25], first, we orient both the detector HWPs at 0°. In this setting, N_{AB} coincidence counts correspond to detections of $|HH\rangle$ photons, and on the other hand $N_{A'B'}$ coincidences correspond to detection of

$|VV\rangle$ photons. Then we modify the orientation of the pump beam HWP until the ratio of the aforementioned coincidence counts is approximately 1:4. We then set the detector HWPs in Alice's and Bob's paths to monitor $N_{AB}(-\alpha, \alpha)$ and carefully tweak the quartz plate tilt to minimize these coincidences. This minimization corresponds to minimizing ϕ in state (5.1). Here, we can clearly see an advantage of using four detectors: when tuning the state, we can monitor $P(-\alpha, \alpha)$ in real time as we make the adjustment in the pump beam polarization.

Once this much has been done, the generated state should be fairly well-tuned according to the requirement. Some final checks would be adjusting the measurement setup to see that the probabilities $P(\beta, \alpha^{\perp})$ and $P(-\alpha^{\perp}, -\beta)$ are both very small (i.e. on the order of a few percent). The state can be further fine-tuned by doing a few iterations of adjusting the pump beam HWP and the quartz plate as detailed in the previous paragraph. Essentially, the goal is to get minimum possible values of $P(-\alpha, \alpha)$, $P(\beta, \alpha^{\perp})$ and $P(-\alpha^{\perp}, -\beta)$ and hence get an increased value of H (see equation (5.51)).

5.5.6 Measuring probabilities with four detectors

If we consider the four-detector measurement setup in figure 5.6, the idler photons move towards Alice and fall on either of her detectors (A and A'). On the other hand, the signal photons move towards Bob and fall on either of his detectors (B and B'). We want to count the photon pairs when Alice's and Bob's detectors register the photons simultaneously. Therefore, like the previous experiment, this experiment is based on counting coincidence events in a certain time period. For instance, $N_{AB'}$ denotes the number of coincidences occurring at the detectors A and B' in a certain counting period.

Once we have measured the coincidence counts represented by N_{AB}, $N_{AB'}$, $N_{A'B}$, and $N_{A'B'}$, we can compute the probability that the photons detected by Alice and Bob had certain polarizations. For instance, let's say that Alice's analysis system is oriented to transmit photons polarized along θ_A to detector A, and Bob's analyzer is oriented to transmit photons polarized along θ_B to detector B. Then, the joint probability of their detecting photons polarized along these directions, as they reach the detectors, $P(\theta_A, \theta_B)$ is given by

$$P(\theta_A, \theta_B) = \frac{N_{AB}}{N_{AB} + N_{AB'} + N_{A'B} + N_{A'B'}}. \tag{5.58}$$

We can write similar expressions for joint probabilities $P(\theta_{A'}, \theta_B)$, $P(\theta_A, \theta_{B'})$ and $P(\theta_{A'}, \theta_{B'})$.

$$P(\theta_A, \theta_{B'}) = \frac{N_{AB'}}{N_{AB} + N_{AB'} + N_{A'B} + N_{A'B'}}, \tag{5.59}$$

$$P(\theta_{A'}, \theta_B) = \frac{N_{A'B}}{N_{AB} + N_{AB'} + N_{A'B} + N_{A'B'}}, \tag{5.60}$$

$$P(\theta_{A'}, \theta_{B'}) = \frac{N_{A'B'}}{N_{AB} + N_{AB'} + N_{A'B} + N_{A'B'}}. \tag{5.61}$$

From now on, for all probabilities represented in the form $P(\theta_A, \theta_B)$, we will take the first angle to correspond to Alice's polarization axis.

If, instead of four-detector measurement setting, we use two-detector setting as depicted in figure 5.1, four different settings of the polarizers are required to measure the coincidences N_{AB}, $N_{AB'}$, $N_{A'B}$, and $N_{A'B'}$.

5.5.7 The experiment

The experimental setup and procedures for aligning the detectors and tuning the Hardy state have been explained in the previous sections. Once this much is accomplished, we are ready to perform the actual nonlocality test. Motorized mounts are used to rotate the wave plates. This can make the experiment convenient to perform.

Photographed in figure 5.7, the experiment is performed using the four-detector scheme. We create the state given by equation (5.57), for which the analysis angles are $\alpha = 35°$, $\beta = 19°$. Since the analyzers comprise HWPs and beam splitters instead of polarizers, the HWPs need to be rotated at half the analysis angles with respect to the horizontal. The coincidence count data are collected for an integration time of two minutes. The required probabilities come out to be

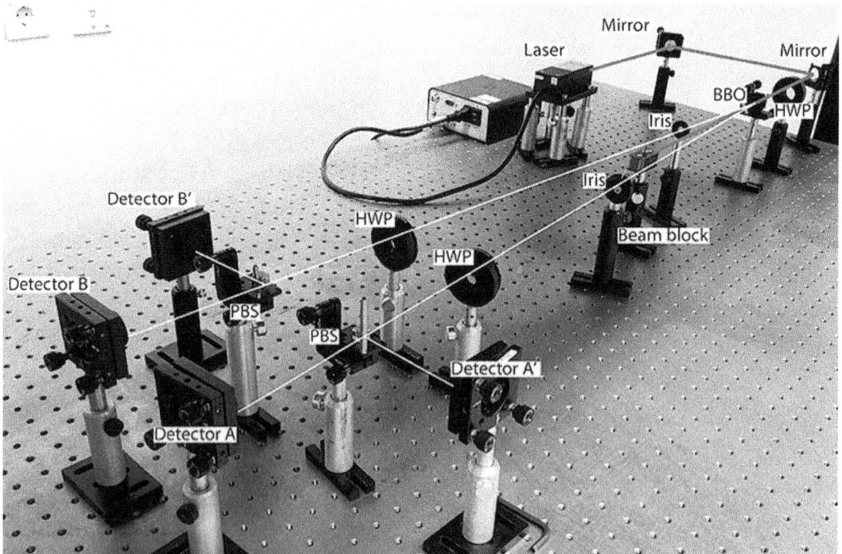

Figure 5.7. Photograph of the Hardy's test setup, excluding the quartz plate.

$$P(\beta, -\beta) = 0.686 \pm 0.002, \tag{5.62a}$$

$$P(\beta, \alpha^{\perp}) = 0.122 \pm 0.001, \tag{5.62b}$$

$$P(-\alpha^{\perp}, -\beta) = 0.238 \pm 0.001, \text{ and} \tag{5.62c}$$

$$P(-\alpha, \alpha) = 0.198 \pm 0.001, \tag{5.62d}$$

and ultimately we determine $H = 0.127 \pm 0.003$. This is a conclusive violation of local realism. The experimental result violates the inequality $H \leqslant 0$ by 44 standard deviations.

5.6 NL3: CHSH test of local realism

The nonlocality test due to CHSH is a significant test of local realism and has been around longer than Hardy's and Freedman's tests [31], though it is more laborious as far as the number of measurements are concerned. As hinted before, the setup is exactly the same as that of Hardy's and can be of two variants: two-detector and four-detector. Four-detector setup reduces the number of trials for one complete experiment, which is why prefer it. Let us first look at the inequality from a purely conceptual perspective.

5.6.1 The CHSH inequality

To derive this inequality, we follow the approach presented in [17]. Consider once again the Alice–Bob experiment of figure 5.1. For a hidden variable theory with hidden variable denoted as λ, we can posit the probability distribution

$$p(\lambda) \geqslant 0, \tag{5.63}$$

which of course is normalized,

$$\int p(\lambda) d\lambda = 1. \tag{5.64}$$

Locality and reality assumptions are included in the argument as follows. For a photon going to Alice, the measurement outcome is completely determined by $A(\alpha, \lambda)$, which is a function of the measurement angle α and the hidden variable λ. Its value is taken to be -1 for H_α and $+1$ for V_α, where H_α and V_α mean Alice's photon is measured in $\{|H\rangle, |V\rangle\}$ basis rotated at an angle α with respect to the horizontal. Similarly, Bob's photon measurement can be described by $B(\beta, \lambda)$. It takes the value of -1 for H_β and $+1$ for V_β, where H_β and V_β mean Bob's photon is measured in a β-rotated HV basis. A hidden variable theory with hidden variable λ therefore specifies $p(\lambda)$, $A(\alpha, \lambda)$ and $B(\beta, \lambda)$.

We know that the probability of a specific outcome, averaged over an ensemble of photon pairs, is given by an integral. Therefore, the probabilities of finding photon pairs in the polarization combinations $V_\alpha V_\beta$, $V_\alpha H_\beta$, $H_\alpha V_\beta$ and $H_\alpha H_\beta$ are given by the following expressions respectively

$$P_{VV}(\alpha, \beta) = \int \left(\frac{1 + A(\alpha, \lambda)}{2}\right)\left(\frac{1 + B(\beta, \lambda)}{2}\right) p(\lambda)d\lambda, \tag{5.65}$$

$$P_{VH}(\alpha, \beta) = \int \left(\frac{1 + A(\alpha, \lambda)}{2}\right)\left(\frac{1 - B(\beta, \lambda)}{2}\right) p(\lambda)d\lambda, \tag{5.66}$$

$$P_{HV}(\alpha, \beta) = \int \left(\frac{1 - A(\alpha, \lambda)}{2}\right)\left(\frac{1 + B(\beta, \lambda)}{2}\right) p(\lambda)d\lambda, \tag{5.67}$$

$$P_{HH}(\alpha, \beta) = \int \left(\frac{1 - A(\alpha, \lambda)}{2}\right)\left(\frac{1 - B(\beta, \lambda)}{2}\right) p(\lambda)d\lambda. \tag{5.68}$$

The terms of the form $(1 \pm A)/2$ and $(1 \pm B)/2$ modify the overall contribution of the terms $A(\alpha, \lambda)$ and $B(\beta, \lambda)$ from $\{-1, 1\}$ to $\{0, 1\}$. Moreover, they will shortly prove useful to get a convenient expression for a function denoted as $E(\alpha, \beta)$.

To make sense of the expressions in equations (5.65)–(5.68) intuitively, consider the integral for $P_{VV}(\alpha, \beta)$ which determines the probability of detecting Alice's and Bob's photons polarized along V_α and V_β respectively. For a specific value of λ, Alice's term $(1 + A(\alpha, \lambda))/2$ will be equal to 0 or 1 corresponding to whether her photon is polarized along H_α or V_α. Similar explanation goes for Bob's term $(1 + B(\beta, \lambda))/2$. Both terms will only contribute to the overall probability when Alice's photon is polarized along V_α and Bob's photon is polarized along V_β. Hence the integral calculates $P_{VV}(\alpha, \beta)$. We can look at the other integrals likewise.

Using equations (5.65)–(5.68), we can show that

$$E(\alpha, \beta) \equiv P_{HH} + P_{VV} - P_{HV} - P_{VH} = \int A(\alpha, \lambda)B(\beta, \lambda)p(\lambda)d\lambda. \tag{5.69}$$

The variable $E(\alpha, \beta)$ is in fact the expected outcome of local realistic measurements that determine Alice's photon to be polarized along α and Bob's along β. Let us also define s that tells us about the polarization correlation in one pair of photons in terms of the four angles a, a', b and b' [17]:

$$\begin{aligned} s &= A(a, \lambda)B(b, \lambda) - A(a, \lambda)B(b', \lambda) + A(a', \lambda)B(b, \lambda) + A(a', \lambda,)B(b', \lambda) \\ &= A(a, \lambda)(B(b, \lambda) - B(b', \lambda)) + A(a', \lambda)(B(b, \lambda,) + B(b', \lambda,)). \end{aligned} \tag{5.70}$$

It is not hard to see that s can only take the values +2 or −2. If we have an ensemble of photons, the average of s can be calculated as

$$\begin{aligned} S(a, a', b, b') \equiv \langle s \rangle &= \int s(\lambda, a, a', b, b')p(\lambda)d(\lambda) \\ &= E(a, b) - E(a, b') + E(a', b) + E(a', b'). \end{aligned} \tag{5.71}$$

Since s can have only the values ± 2, S which is its average must satisfy

$$|S| \leqslant 2. \tag{5.72}$$

This is called the CHSH inequality and it must be obeyed by any local realistic theory. Next, we look at the quantum mechanical prediction.

5.6.2 Quantum mechanical prediction for the CHSH test

Consider the bipartite Bell state (5.25) reproduced here:

$$|\psi\rangle = \frac{1}{\sqrt{2}}(|H\rangle_A|H\rangle_B + |V\rangle_A|V\rangle_B). \tag{5.73}$$

We revisit the rotated $\{|H\rangle, |V\rangle\}$ basis that we alluded to in the previous section. If we measure polarization in the $\{|H\rangle, |V\rangle\}$ basis that is rotated anticlockwise at an angle α with respect to the horizontal, the basis states are given by

$$|H_\alpha\rangle = \cos\alpha|H\rangle + \sin\alpha|V\rangle, \tag{5.74}$$

$$|V_\alpha\rangle = -\sin\alpha|H\rangle + \cos\alpha|V\rangle. \tag{5.75}$$

The measurement probabilities are then given by

$$P_{VV}(\alpha, \beta) = |(\langle V_\alpha|_A\langle V_\beta|_B)|\psi\rangle|^2 = \frac{1}{2}\cos^2(\alpha - \beta). \tag{5.76}$$

Similarly, we have

$$P_{HH}(\alpha, \beta) = \frac{1}{2}\cos^2(\alpha - \beta) \text{ and} \tag{5.77}$$

$$P_{HV}(\alpha, \beta) = P_{VH}(\alpha, \beta) = \frac{1}{2}\sin^2(\alpha - \beta). \tag{5.78}$$

Plugging equations (5.76)–(5.78) into equation (5.69), we obtain

$$E(\alpha, \beta) = P_{HH} + P_{VV} - P_{HV} - P_{VH} = \cos(2(\alpha - \beta)). \tag{5.79}$$

Using the above equation with equation (5.71), we ultimately get the following expression for S:

$$\begin{aligned} S &= E(a, b) - E(a, b') + E(a', b) + E(a', b') \\ &= \cos(2(a - b)) - \cos(2(a - b')) + \cos(2(a' - b)) \\ &\quad + \cos(2(a' - b')). \end{aligned} \tag{5.80}$$

It can be shown that for angles $a = -45°$, $a' = 0°$, $b = -22.5°$ and $b' = 22.5°$, we get $S = 2\sqrt{2}$. This numerical figure provides the maximal possible violation of CHSH inequality. Other analysis angles may indeed be chosen instead but we will get maximal violation for the angles whose relative disposition is shown in figure 5.8 [17].

It is important to mention here that $S = 2\sqrt{2}$ only for the Bell state given by equation (5.73). Other states may produce lower values of S. Experimentally, it is very hard to perfectly produce the state represented by equation (5.73). Therefore,

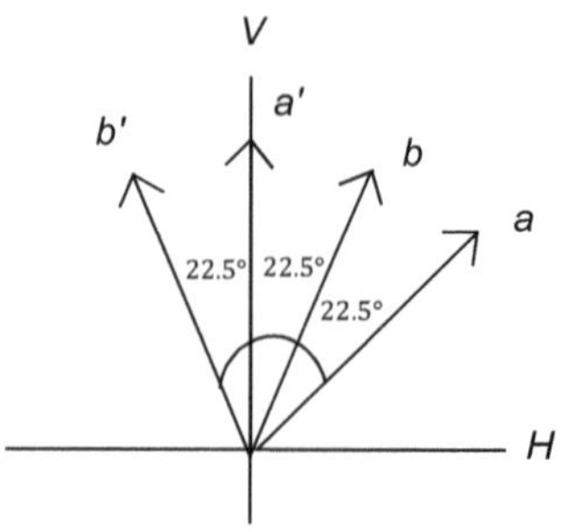

Figure 5.8. Relation between analysis angles for achieving maximal violation of CHSH inequality. According to equation (5.80), the absolute values of the angles do not matter—only the relative orientations do.

we expect to obtain nonideal results. Nevertheless, if we can experimentally show that $S > 2$, it will be a conclusive violation of the Bell's inequality, negating all hidden variable theories.

5.6.3 Tuning the Bell state

Since, the experimental setup for CHSH test is identical to that of Hardy's test, experiment NL2 can be consulted for alignment procedures. However, once the setup is aligned, it is important to tune the Bell state of equation (5.73) in order to achieve maximal violation of the CHSH inequality.

We will create the same Bell state as was used in Freedman's test in experiment NL1. The procedure is analogous but adapted for the four-detector scheme of the CHSH test. Consider the schematic diagram shown in figure 5.6. To create the Bell state, we need to equalize AB and $A'B'$ counts and minimize $A'B$ and AB' counts in both the $\{|H\rangle, |V\rangle\}$ and $\{|D\rangle, |A\rangle\}$ bases. Equalizing AB and $A'B'$ counts essentially corresponds to equalizing A and B in state (5.1) while minimizing $A'B$ and AB' counts *in both bases* means making ϕ equal to zero.

First, we create the $\{|H\rangle, |V\rangle\}$ measurement basis. The A and B HWP angles are set to 0°. With this arrangement, the A and B detectors register horizontally polarized photons generated by the source, and the A' and B' detectors register vertically polarized photons. We adjust the pump beam HWP until the AB and $A'B'$ coincidences are maximized and their ratio is roughly 1:1. The $A'B$ and AB' coincidences should, however, be minimum.

Now we change the measurement basis to $\{|D\rangle, |A\rangle\}$ by setting the A and B HWPs to 22.5°, and tweak the tilt of the quartz plate to minimize the $A'B$ and AB' coincidences. The AB and $A'B'$ coincidences should still be maximum and approximately in equal number.

To fine-tune the state, we perform a few iterations of the aforementioned two steps. Ultimately, when AB and $A'B'$ coincidences are roughly in the ratio 1:1 and $A'B$ and AB' coincidences are minimum in both measurement bases, the downconverted photons are approximately generated in the required Bell state. We share the experimental results of the CHSH test in the next section.

Table 5.2. Results of the CHSH test of local realism. The column labeled 'Confidence' lists the number of standard deviations σ by which the experimental results violate $|S| \leqslant 2$. The violation from the local realistic theories becomes stronger for longer integration times.

Integration time	S	Confidence
20 s	2.504 ± 0.015	33 σ
20 s	2.568 ± 0.013	44 σ
120 s	2.454 ± 0.006	69 σ

5.6.4 The experiment

The schematic of this experiment can be seen in figure 5.6 while a photograph is shown in figure 5.7. Following the procedure in the previous section, the Bell state is produced, for which the analysis angles are chosen to be $a = -45°$, $a' = 0°$, $b = -22.5°$ and $b' = 22.5°$.

Since we are using the four-detector scheme with HWPs and beam splitters instead of polarizers, we orient the detector HWPs to half the analysis angles and then detect coincidence counts for a certain integration time. We record three datasets and summarize the experimental results in table 5.2. These results clearly show that the system under study violates local realism, agreeing with the quantum mechanical prediction ($S > 2$).

Arguably, one of the most attractive aspects of physics is pondering over the nature of physical reality. Many different Bell tests have been performed and re-performed to this day, each version seemingly better than the previous ones and closing more loopholes. We have discussed three very accessible versions of Bell tests in this chapter. Once the experimental setups are ready, these experiments can be performed in a single afternoon. Yet it seems that these experiments may be one of the deepest investigations of the ultimate nature of reality.

References

[1] Schrödinger E 1935 Discussion of probability relations between separated systems *Mathematical Proceedings of the Cambridge Philosophical Society* vol 31 (Cambridge: Cambridge University Press) pp 555–63
[2] Einstein A, Podolsky B and Rosen N 1935 *Phys. Rev.* **47** 777
[3] Ekert A K 1991 *Phys. Rev. Lett.* **67** 661
[4] Ekert A K, Rarity J G, Tapster P R and Palma G M 1992 *Phys. Rev. Lett.* **69** 1293
[5] Bennett C H *et al* 1993 *Phys. Rev. Lett.* **70** 1895
[6] Bennett C H and Wiesner S J 1992 *Phys. Rev. Lett.* **69** 2881
[7] Hidary J D 2019 *Quantum Computing: An Applied Approach* (Berlin: Springer)
[8] Stapp H P 1972 *Am. J. Phys.* **40** 1098
[9] Bell J S 1964 *Phys. Phys. Fizika* **1** 195
[10] Bell J S 2004 *Speakable and Unspeakable in Quantum Mechanics: Collected Papers on Quantum Philosophy* (Cambridge: Cambridge University Press)

[11] Freedman S J 1972 Experimental test of local hidden-variable theories *PhD Thesis*
[12] Clauser J F, Horne M A, Shimony A and Holt R A 1969 *Phys. Rev. Lett.* **23** 880
[13] Hardy L 1993 *Phys. Rev. Lett.* **71** 1665
[14] Freedman S J and Clauser J F 1972 *Phys. Rev. Lett.* **28** 938
[15] Pan J-W, Bouwmeester D, Daniell M, Weinfurter H and Zeilinger A 2000 *Nature* **403** 515
[16] Torgerson J R, Branning D, Monken C H and Mandel L 1995 *Phys. Lett.* A **204** 323
[17] Dehlinger D and Mitchell M 2002 *Am. J. Phys.* **70** 903
[18] Galvez E J 2014 *Am. J. Phys.* **82** 1018
[19] Greenberger D M, Horne M A and Zeilinger A 1989 Going beyond Bells theorem *Bell's Theorem, Quantum Theory and Conceptions of the Universe* (Berlin: Springer) pp 69–72
[20] Mermin N D 1990 *Am. J. Phys.* **58** 731
[21] Mermin N D 1995 *Ann. N. Y. Acad. Sci.* **755** 616
[22] Di Giuseppe G, De Martini F and Boschi D 1997 *Phys. Rev.* A **56** 176
[23] White A G, James D F, Eberhard P H and Kwiat P G 1999 *Phys. Rev. Lett.* **83** 3103
[24] Brody J and Selton C 2018 *Am. J. Phys.* **86** 412
[25] Carlson J, Olmstead M and Beck M 2006 *Am. J. Phys.* **74** 180
[26] Mermin N D 1994 *Am. J. Phys.* **62** 880
[27] Kwiat P G and Hardy L 2000 *Am. J. Phys.* **68** 33
[28] Branning D 1997 *Am. Sci.* **85** 160
[29] Kwiat P G, Waks E, White A G, Appelbaum I and Eberhard P H 1999 *Phys. Rev.* A **60** R773
[30] Dehlinger D and Mitchell M 2002 *Am. J. Phys.* **70** 898
[31] Beck M 2012 *Quantum Mechanics: Theory and Experiment* (Oxford: Oxford University Press)
[32] Aspect A, Grangier P and Roger G 1981 *Phys. Rev. Lett.* **47** 460
[33] Clauser J F and Horne M A 1974 *Phys. Rev.* D **10** 526
[34] Hensen B *et al* 2015 *Nature* **526** 682
[35] Giustina M *et al* 2015 *Phys. Rev. Lett.* **115** 250401
[36] Shalm L K *et al* 2015 *Phys. Rev. Lett.* **115** 250402
[37] Larsson J-Å 2014 *J. Phys. A Math. Theor.* **47** 424003

Muhammad Hamza Waseem, Faizan-e-Ilahi and Muhammad Sabieh Anwar

Chapter 6

Quantum state tomography

In experiments Q3 and Q4, we used one of the two downconverted photons as a herald for detection of the other photon. In other words, our focus was on studying the quantum state of the signal photon. The complete study of entangled states, however, requires investigating the quantum state of *both* the photons forming the entangled pair. On the other hand, in experiments NL1–NL3, we saw that quantum mechanics explains certain two-photon polarization states which violate the apparently self-evident assumptions of locality and/or reality. This pair of photons was *entangled* in polarization.

In this chapter, we will attempt to account for the polarization-based description of the *complete* pair of downconverted photons. This is achieved through the technique of quantum state tomography (QST).

Quantum state tomography is a fancy name for estimating the quantum state. It is achieved by using a number of carefully orchestrated measurements. There are a few caveats though. As established in experiments Q3 and Q4, performing measurement of a quantum particle perturbs its state. Therefore, quantum state tomography cannot be used to unambiguously determine the state of a *single* particle in just one go. Rather, it is performed systematically in successive stages on many identical copies of the quantum state under investigation and since there can never be an infinite supply of particles, so we can only make educated best guesses about the state. In all cases, we need to devise an algorithmic procedure for this state estimation.

We have seen in chapter 2 that the Stokes parameters S_0, S_1, S_2 and S_3 describe the polarization state of light [1]. These parameters completely characterize the state. One can therefore imagine that state tomography entails measurement of precisely these parameters. This is indeed true.

We have also learned that certain light beams can be seen as ensembles of two-level quantum systems (comprising photons), with the two polarization degrees of freedom defining the quantum state. The polarization state is completely described by a density matrix of the form

doi:10.1088/978-0-7503-3063-3ch6

$$\hat{\rho} = \begin{pmatrix} a & b + ic \\ b - ic & 1 - a \end{pmatrix} \quad (6.1)$$

where a, b and c are real numbers making $\hat{\rho}$ into a unit trace and Hermitian, positive semi-definite matrix. There are three real parameters, the a, b and c that need to be estimated if the state is to be characterized. This apparently contradicts the need for specifying four Stokes parameters. This discrepancy is resolved by noting that the parameter S_0 depends on the intensity in the classical case and the number of counts in the quantum mechanical case. Measuring for longer times over higher intensity beams will increase S_0, which does not really affect the polarization state, but merely distinguishes how big the observed signal is. Therefore one of the Stokes parameters is determining the signal-to-noise ratio and it is only the trio S_1, S_2 and S_3 normalized against S_0 which characterizes the state. Therefore, there should exist a direct one-to-one mapping between the Stokes parameters and the real parameters a, b and c that describe the density matrix. This chapter will build upon this understanding of polarization states.

There exist simple tomographic techniques in which experimental data are linearly transformed to find the density matrix of a quantum state. However, because of experimental noise, these methods may sometimes not fetch density matrices corresponding to realizable physical states describable by Hermitian, positive semi-definite matrices of unit trace of the form (6.1). In fact, many experimentally measured matrices usually fail to fulfill the positive semi-definiteness condition [2]. As a work-around, a maximum likelihood estimation approach has been proposed and successfully utilized [3–7]. According to this method, the density matrix which has the maximum likelihood of having generated the measured dataset is obtained via numerical optimization and the semi-definite positivity condition is hard-wired into the optimization routine. This chapter's discussions closely follow [8] which is an excellent introduction to using maximum likelihood techniques for estimating polarization states of photons.

A little bit of survey here. Tomographic methods have been successfully used for measurement in quantum mechanical systems of a vast variety and complexity [9–15]. Particularly, these methods have worked remarkably well in studying the quantum state of qubits based on polarization-entangled pairs of photons generated through downconversion [16]. Our focus will be on the QST of such systems, but the discussion is applicable to other two-level quantum states as well.

In this chapter, we will extend the discussion of techniques for QST, already initiated in experiments Q3 and Q4, in the context of correlated two-level quantum optical systems or qubits. We will particularly focus on two techniques, namely a linear tomographic reconstruction and a maximum likelihood technique. The former technique linearly relates the density matrix to a series of measurements and is important for understanding the theoretical basis of tomography using idealized measurements. However, the method does not always return valid density matrices for physical systems. On the other hand, the latter technique is based on numerical optimization and is an adaptation of the former technique to real, nonideal systems. Hence, it returns valid density matrices for imperfect experimental conditions as well. This is the method we have employed for estimating the quantum state in our experiments. To begin our discussion, we revisit the relation between the Stokes parameters and QST.

6.1 Qubits, Stokes parameters and tomography

As mentioned in the introduction, an analogy exists between measuring the polarization state of light and estimating the density matrix of two-level quantum systems. This analogy is the subject of this section. Before discussing tomography for two qubits, we will investigate the single-qubit case. This will help us develop an intuitive picture of quantum state representation and tomography before we can investigate the extended case of two qubits.

6.1.1 The Bloch sphere for pure states

As we saw in chapter 4, we can generally express any single qubit by equation (4.28) if it is a pure state. This representation is not only sufficient for describing a pure state but also allows us to describe the action of an operator by straightforward matrix manipulation. For example, projection or unitary operation on the pure state $|\psi\rangle$ can be described by $\hat{P}_\phi|\psi\rangle = |\phi\rangle\langle\phi|\psi\rangle$ and $\hat{U}|\psi\rangle$ respectively.

Figure 6.1(a) shows a pure state $|\psi\rangle$ lying on the surface of the Bloch sphere. The sphere identifies three axes as pointing along the pure states $|D\rangle$, $|L\rangle$ and $|H\rangle$. Earlier, we have also used different labels for these states:

$$|H\rangle \equiv |0\rangle \tag{6.2}$$

$$|L\rangle \equiv \frac{1}{\sqrt{2}}(|0\rangle + i|1\rangle), \quad \text{and} \tag{6.3}$$

$$|D\rangle \equiv \frac{1}{\sqrt{2}}(|0\rangle + |1\rangle). \tag{6.4}$$

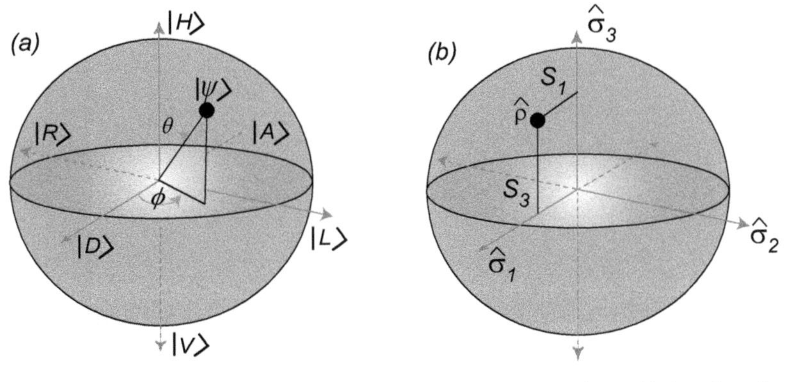

Figure 6.1. (a) Any pure qubit state $|\psi\rangle$ can be represented on the Bloch sphere. (b) Another picture is useful when we consider general mixed states. This latter is also the Bloch sphere with the axes relabeled as the three zero-trace Pauli operators.

Each axis represents one choice of the basis vectors. For example the axis pointing in the $|D\rangle$ direction corresponds to the $\{\mathbf{e}_D, \mathbf{e}_A\}$ basis; $|L\rangle$ defines the $\{\mathbf{e}_L, \mathbf{e}_R\}$ basis and finally $|H\rangle$ defines the orientation of the $\{\mathbf{e}_H, \mathbf{e}_V\}$ basis. Note that quantum mechanically, $|H\rangle$ and $|D\rangle$ are not orthogonal ($\langle D|H\rangle = 1\sqrt{2}$) but in the Bloch sphere picture, these states lie along orthogonal axes. Quantum orthogonality does not mean orthogonality in the Bloch sphere representation.

The pure state $|\psi\rangle$ is a point on the surface and its position is characterized by angles θ and ϕ. In fact, these angles act as parameters that map any pure state onto the surface of the sphere and are shown in figure 6.1(a). These values are the polar coordinates of the pure state they represent. Apart from representation of a single-qubit state, the Bloch sphere serves another purpose too. It can be used to represent a unitary operation. On the Bloch sphere, any unitary operation is mapped to a rotation about an arbitrary axis. We have seen that waveplates are used to perform unitary operations and hence, in the Bloch sphere picture, act as rotations about some axis. The axis of rotation is affixed by the orientation of the fast axis of the wave plate and the amount of rotation by the kind of wave plate (half wave or full wave).

6.1.2 The Bloch sphere for density matrices

If the to-be-measured ensemble of particles forms a mixed state, we need to resort to density matrices, which can generally be expressed as

$$\hat{\rho} = \sum_i p_i |\psi_i\rangle\langle\psi_i| \tag{6.5}$$

$$= \begin{pmatrix} A & Ce^{i\phi} \\ Ce^{-i\phi} & B \end{pmatrix}, \tag{6.6}$$

which is just another way of writing down the matrix in equation (6.1). Here A, B, and C are real, non-negative numbers, and are related as $A + B = 1$ and $C = \sqrt{AB}$. The coefficient p_i denotes the probabilistic weighting ($\sum_i p_i = 1$) [17]. Completely determining this matrix requires $4 - 1 = 3$ parameters (since $A + B = 1$). The relationship between these parameters is as follows:

$$A = a, \tag{6.7}$$

$$B = 1 - A = 1 - a, \tag{6.8}$$

$$b = C\cos\phi, \tag{6.9}$$

$$c = C\sin\phi. \tag{6.10}$$

Using the definitions of the canonical polarization states spelled out in table 3.1, we can construct the so-called Pauli spin operators as

$$\hat{\sigma}_1 \equiv |D\rangle\langle D| - |A\rangle\langle A| = \begin{pmatrix} 0 & 1 \\ 1 & 0 \end{pmatrix}, \tag{6.11}$$

$$\hat{\sigma}_2 \equiv |L\rangle\langle L| - |R\rangle\langle R| = \begin{pmatrix} 0 & -i \\ i & 0 \end{pmatrix}, \tag{6.12}$$

$$\hat{\sigma}_3 \equiv |H\rangle\langle H| - |V\rangle\langle V| = \begin{pmatrix} 1 & 0 \\ 0 & -1 \end{pmatrix} \tag{6.13}$$

and the 2×2 identity matrix

$$\hat{\sigma}_0 \equiv |H\rangle\langle H| + |V\rangle\langle V| = \begin{pmatrix} 1 & 0 \\ 0 & 1 \end{pmatrix}. \tag{6.14}$$

These operators $\{\hat{\sigma}_0, \hat{\sigma}_1, \hat{\sigma}_2, \hat{\sigma}_3\}$ form a complete basis for 2×2 Hermitian matrices and hence the density matrix in equation (6.5) can also be expressed as

$$\hat{\rho} = \frac{1}{2}\sum_{i=0}^{3} S_i \hat{\sigma}_i, \tag{6.15}$$

where $(S_1, S_2, S_3)^T$ is sometimes called the *Bloch vector* or the Stokes vector. The coefficients are simply the Stokes parameters for light. We will show that in a moment. Due to the unit trace condition $S_0 = 1$ always but we keep on writing it for convenience.

A more general view of the Bloch sphere is illustrated in figure 6.1(b) which relabels the axes as the Pauli operators. This conceptual diagram allows us to represent mixed states in addition to pure states. Consider a quantum state (6.15) which lives somewhere on the Bloch sphere. In chapter 3, we have learned that for a pure state $\hat{\rho}^2 = \hat{\rho}$, so we have $\mathrm{Tr}(\hat{\rho}^2) = 1$. Using the representation of the density matrix above, we have

$$\hat{\rho}^2 = \frac{1}{4}\left(\sum_{i=0}^{3} S_i \hat{\sigma}_i\right)\left(\sum_{k=0}^{3} S_k \hat{\sigma}_k\right) \tag{6.16}$$

$$= \frac{1}{4}\sum_{i,k=0}^{3} S_i S_k \hat{\sigma}_i \hat{\sigma}_k \tag{6.17}$$

which leads to (using $S_0 = 1$)

$$\mathrm{Tr}(\hat{\rho}^2) = \frac{1}{2}\left(1 + S_1^2 + S_2^2 + S_3^2\right). \tag{6.18}$$

Therefore pure states have $S_1^2 + S_2^2 + S_3^2 = 1$ and for mixed states, $S_1^2 + S_2^2 + S_3^2 < 1$ with the extreme case $S_1^2 + S_2^2 + S_3^2 = 0$ denoting the maximally mixed state.

This discussion reveals an interesting geometrical interpretation of the S coefficients. The mixed state $\hat{\rho}$ in figure 6.1(b) is shown to lie inside the Bloch sphere which is now shown with the relabeled axes. Suppose the state is represented by the solid circle and lies wholly in the $\hat{\sigma}_1$–$\hat{\sigma}_3$ plane. Therefore one can imagine that while part (a) of this figure shows states inside the Hilbert space, part (b) shows states inside the operator space[1]. If we were to drop a projection of the state onto the $\hat{\sigma}_1$ axis, we can measure the distance from the origin. This distance is measured to be S_1. Similarly if the projection is taken along the $\hat{\sigma}_2$ axis, the distance is S_2. For our particular chosen state, $S_2 = 0$. Finally, the distance of the projected point along the $\hat{\sigma}_3$ axis is S_3. Each of these S parameters can be positive or negative; positive when the projection is on one side of the origin and negative when on the diametrically opposite side. An example of a state with a negative value of S_1 is shown in figure 6.2. We now motivate why these coefficients are the familiar Stokes parameters.

6.1.3 Stokes parameters as projections of the state on the Bloch sphere

The Stokes parameters, which provide a complete description of the polarization state of light, are classically defined in terms of intensity measurements in the $\{\mathbf{e}_D, \mathbf{e}_A\}$, $\{\mathbf{e}_L, \mathbf{e}_R\}$ and $\{\mathbf{e}_H, \mathbf{e}_V\}$ bases [18]. For single photons, the intensities can be replaced by numbers of photodetections, or probabilities which are directly proportional to photocounts. If the basis states are given by $\{|D\rangle, |A\rangle\}$, $\{|L\rangle, |R\rangle\}$ and $\{|H\rangle, |V\rangle\}$ representing the canonical polarization states (see table 3.1 in chapter 3), the Stokes parameters are defined as [8]

$$S_0 = P_{|H\rangle} + P_{|V\rangle}, \tag{6.19a}$$

$$S_1 = P_{|D\rangle} - P_{|A\rangle}, \tag{6.19b}$$

$$S_2 = P_{|L\rangle} - P_{|R\rangle}, \tag{6.19c}$$

$$S_3 = P_{|H\rangle} - P_{|V\rangle}, \tag{6.19d}$$

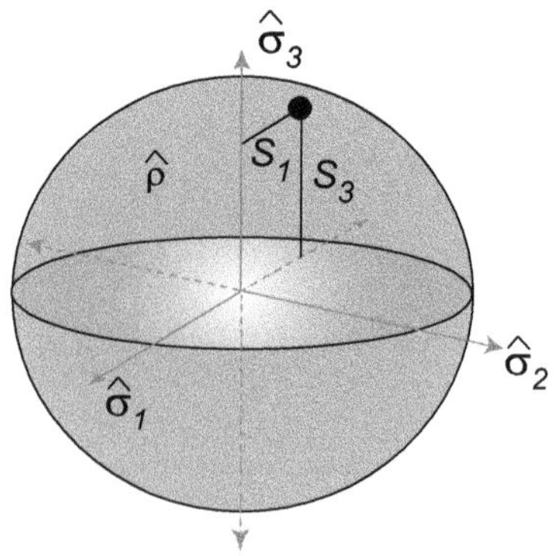

Figure 6.2. A mixed state with a negative value for the parameter S_1.

[1] Sometimes this is called the Liouville space.

where $P_{|\phi\rangle}$ is the probability of measuring the photon in the polarization state $|\phi\rangle$. In chapter 2 which dealt with a purely classical description of light, we had defined instead the Stokes parameters as

$$S_0 = I_H + I_V, \tag{6.20a}$$

$$S_1 = I_D - I_A, \tag{6.20b}$$

$$S_2 = I_L - I_R, \tag{6.20c}$$

$$S_3 = I_H - I_V, \tag{6.20d}$$

with the I_H representing the intensity of detecting the light with horizontal (H) polarization etc. Comparing the eight definitions just stated, we see that the quantum version of Stokes parameters is inferred from probabilities whereas in the classical picture, they are inferred from intensities. In quantum experiments, probabilities are deduced from normalized counts whereas in classical experiments, intensities arise out of photocurrents spewed out by detectors.

It is also important to remember that for a state described by the density matrix $\hat{\rho}$, $P_{|\phi\rangle}$ is given by

$$P_{|\phi\rangle} = \langle\phi|\hat{\rho}|\phi\rangle = \mathrm{Tr}\{|\phi\rangle\langle\phi|\hat{\rho}\}. \tag{6.21}$$

Using equations (6.5) and (6.21), we have

$$\begin{aligned} P_{|\phi\rangle} &= \langle\phi|\left(\sum_i p_i |\psi_i\rangle\langle\psi_i|\right)|\phi\rangle \\ &= \sum_i p_i \langle\phi|\psi_i\rangle\langle\psi_i|\phi\rangle \\ &= \sum_i p_i |\langle\phi|\psi_i\rangle|^2 \end{aligned} \tag{6.22}$$

where each term $|\langle\phi|\psi_i\rangle|^2$ represents the probability of projecting state $|\psi_i\rangle$ onto $|\phi\rangle$. These terms are mixed up the weighting factor p_i which shows what fraction of $|\psi_i\rangle\langle\psi_i|$ contributes to the statistical mixture $\hat{\rho}$.

Using the definition of the state (6.15) and how we extract probabilities from it, equation (6.21), we obtain the series of straightforward results

$$S_1 = \mathrm{Tr}(\hat{\rho}\hat{\sigma}_1) = \mathrm{Tr}\,(\hat{\rho}(|D\rangle\langle D| - |A\rangle\langle A|)) = P_{|D\rangle} - P_{|A\rangle} \tag{6.23}$$

$$S_2 = \mathrm{Tr}(\hat{\rho}\hat{\sigma}_2) = \mathrm{Tr}\,(\hat{\rho}(|L\rangle\langle L| - |R\rangle\langle R|)) = P_{|L\rangle} - P_{|R\rangle} \tag{6.24}$$

$$S_3 = \mathrm{Tr}(\hat{\rho}\hat{\sigma}_3) = \mathrm{Tr}\,(\hat{\rho}(|H\rangle\langle H| - |V\rangle\langle V|)) = P_{|H\rangle} - P_{|V\rangle}. \tag{6.25}$$

These results show that the coefficients in the density matrix truly *coincide* with the Stokes parameters. The act of determining these coefficients is tantamount to Stokes polarimetry. In short, the goal of QST is precisely to determine the Stokes parameters and these allow us to reconstruct the complete density matrix.

Equations (6.15), (6.19), (6.23)–(6.25) form the perfect recipe for reconstructing density matrices of single-qubit states as long as no errors and perfect experimental conditions are assumed. We will illustrate their use in the following section.

6.2 Single-qubit tomography

As stated earlier, state tomography is aimed at estimating the density matrix of an unknown ensemble of particles via a series of carefully planned measurements. In the absolute sense, this cannot be performed exactly, because it requires an infinitely many number of particles and measurements to eliminate the statistical error.

Though it is unrealistic, we will assume no sources of error in this section and still perform single-qubit tomography. This exercise will employ the toolbox that we have seen in the previous section. Later in this chapter, we will explore two-qubit tomography while considering sources of errors and their compensation.

Assuming perfectly exact measurements, single-qubit state tomography requires making sets of three linearly independent measurements. Let's see what this means. From equations (6.23)–(6.25), we understand that we require to measure *six* probabilities, $P_{|D\rangle}$, $P_{|A\rangle}$, $P_{|L\rangle}$, $P_{|R\rangle}$, $P_{|H\rangle}$ and $P_{|V\rangle}$. There are several ways in which this can be achieved. We describe only the approach aligned to what we have actually employed in our experiments.

Suppose we have *two* detectors and *three* different kinds of beam splitters at our disposal. These are shown in figure 6.3. Part (a) shows a beam splitter that routes the input photon into possible output channels depending on the polarization state $\hat{\rho}$ of the input beam. One is the $|H\rangle$ channel and the other is the $|V\rangle$ channel. Let's call this an *HV* beam splitter. The photons emerging from the channels can be counted with the help of single photon counting modules. The normalized counts determine the probabilities $P_{|H\rangle}$ and $P_{|V\rangle}$ from which S_3 (as well as $S_0 = P_{|H\rangle} + P_{|V\rangle}$) can be determined. In terms of the Bloch sphere picture, this helps determine how high or low the state $\hat{\rho}$ is with respect to the $\hat{\sigma}_2$–$\hat{\sigma}_3$ plane. The measurement thus collapses the unknown state onto a plane parallel to $\hat{\sigma}_2$–$\hat{\sigma}_3$. We perform these measurements on an ensemble of identical single photons. No doubt, there should be a large enough number to allow the counts approach the expected values.

We then switch to another beam splitter, say the *DA* beam splitter which routes the state into $|D\rangle$ and $|A\rangle$ channels. This is shown in figure 6.3(b). These measurements yield $P_{|D\rangle}$ and $P_{|A\rangle}$ allowing us to deduce S_1, collapsing the possible planar

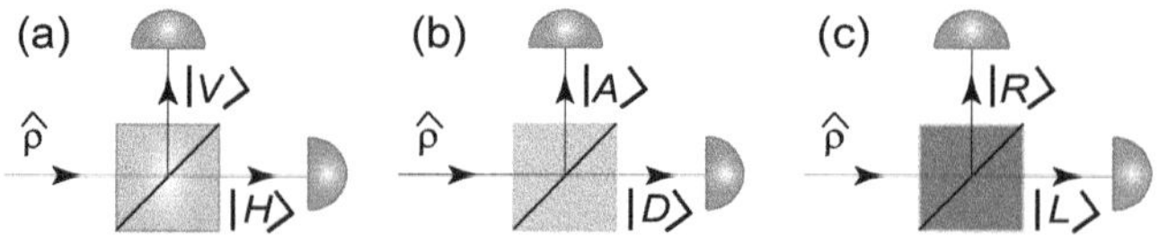

Figure 6.3. Three settings with two detectors gives six different kinds of measurements. The subfigures (a)–(c) are conceptual measurement schemes using three different kinds of beam splitters, the *HV*, *DA* and *LR* beam splitters respectively.

solution to a line parallel to the $\hat{\sigma}_1$ axis. Last, measurement with the *LR* beam splitter determines S_2 pinpointing a single point (of course with some uncertainty) on the line extracted from the previous experiment.

In summary, before the measurement, the unknown state may be anywhere on or inside the Bloch sphere—we do not know. The first set of measurements always caps the state under investigation to a plane. The second set of measurements isolates the unknown state from the plane to a line. From this line, the third set of measurements finally fetches a point on the Bloch sphere, corresponding to an estimate of the measurand state.

Let's recount the physical resources required for this tomographic procedure. We have three settings corresponding to three different kinds of experimental arrangements (in this case three different kinds of physically distinct beam splitters). We make two measurements within each setting, taking our total measurements number to six. From these six outcomes, we deduce the Stokes parameters which allow us to estimate the density matrix.

But hold on, there is a problem. In the laboratory, one may not find these three kinds of rather unusual beam splitters. The *HV* beam splitter is a common lab item but achieving projections in the $\{|D\rangle, |A\rangle\}$ and $\{|L\rangle, |R\rangle\}$ bases is not clear. In order to bypass this, one can follow the measurement scheme outlined in section 4.4.2 of chapter 4. The scheme involves placing waveplates in front of the *HV* beam splitter and orienting them to create the required measurement basis.

This deserves a bit of an elaboration. The basic philosophy goes as follows. Suppose we need to project the input state along the $\{|D\rangle, |A\rangle\}$ basis. Instead of searching for a beam splitter that can analyze the beam in this fashion, why not rotate the input state so that $|D\rangle$ is rotated to $|H\rangle$ and $|A\rangle$ is rotated to $|V\rangle$ and still keep to the use of *HV* beam splitter. Thus the statistics of the detector placed inside the $|H\rangle$ channel firing for this arrangement will be identical to the statistics achieved by looking at the detector placed inside the $|D\rangle$ channel of a fictitious *DA* beam splitter. Therefore, corresponding to the measurements suggested in figure 6.3(b) and (c), we can use a combination of quarter-wave plate and a half-wave plate oriented at angles q and h. The setup is schematically depicted in figure 6.4. For analysis along the $\{|D\rangle, |A\rangle\}$ basis, we require $q = 45°$, $h = 22.5°$ and for analysis along the $\{|L\rangle, |R\rangle\}$ basis, we use the setting $q = 45°$, $h = 0°$. Furthermore, we can also do the $\{|H\rangle, |V\rangle\}$ analysis with this arrangement by setting $q = 0°$, $h = 0°$. The effective impact of this setup is to rotate the density matrix (living in the Bloch sphere) so that the desired axes coincide, in a systematic fashion, with the canonical axes of the *HV* beam splitter.

Recalling experiment Q3, we generated single-qubit states and performed measurements in the aforementioned bases. Hence, we can use the same measurements to estimate the respective density matrices. The procedure is as follows.

Measurements in the $\{|H\rangle, |V\rangle\}$, $\{|D\rangle, |A\rangle\}$ and $\{|L\rangle, |R\rangle\}$ bases let us determine the set of probabilities $P_{|H\rangle}$, $P_{|V\rangle}$, $P_{|D\rangle}$, $P_{|A\rangle}$, $P_{|L\rangle}$ and $P_{|V\rangle}$. Using these probabilities, we calculate the Stokes parameters using the relations given in equation (6.19). Finally, the density matrices are computed using equation (6.15). The results, generated from measurements made in experiment Q3, are presented in table 6.1.

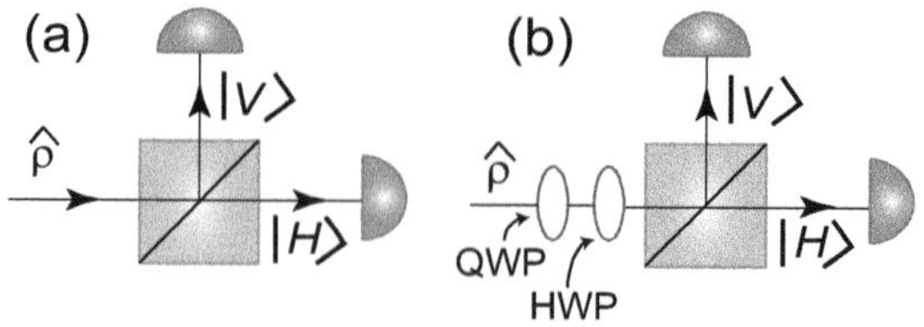

Figure 6.4. The measurement basis can be transformed by adopting an active approach of rotating the input state $\hat{\rho}$: (a) shows how to measure S_3 which is a projection in the $\{|H\rangle, |V\rangle\}$ basis while (b) shows how the placement of properly oriented quarter-wave plate (QWP) and half-wave plate (HWP) in the path of photons can rotate the state so that it reorients with the rotated measurement basis.

Table 6.1. Results of single-qubit tomography.

State	Predicted density matrix	Measured density matrix
$\lvert H\rangle$	$\begin{pmatrix} 1 & 0 \\ 0 & 0 \end{pmatrix}$	$\begin{pmatrix} 0.99 & -0.03 - 0.34i \\ -0.03 + 0.34i & 0.01 \end{pmatrix}$
$\lvert V\rangle$	$\begin{pmatrix} 0 & 0 \\ 0 & 1 \end{pmatrix}$	$\begin{pmatrix} 0.06 & 0.10 + 0.24i \\ 0.10 - 0.24i & 0.94 \end{pmatrix}$
$\lvert D\rangle$	$\begin{pmatrix} 0.5 & 0.5 \\ 0.5 & 0.5 \end{pmatrix}$	$\begin{pmatrix} 0.56 & 0.35 - 0.07i \\ 0.35 + 0.07i & 0.44 \end{pmatrix}$
$\lvert A\rangle$	$\begin{pmatrix} 0.5 & -0.5 \\ -0.5 & 0.5 \end{pmatrix}$	$\begin{pmatrix} 0.49 & -0.45 - 0.04i \\ -0.45 + 0.04i & 0.51 \end{pmatrix}$
$\lvert L\rangle$	$\begin{pmatrix} 0.5 & -0.5i \\ 0.5i & 0.5 \end{pmatrix}$	$\begin{pmatrix} 0.54 & 0.04 - 0.48i \\ 0.04 + 0.48i & 0.46 \end{pmatrix}$
$\lvert R\rangle$	$\begin{pmatrix} 0.5 & 0.5i \\ -0.5i & 0.5 \end{pmatrix}$	$\begin{pmatrix} 0.44 & 0.02 + 0.44i \\ 0.02 - 0.44i & 0.56 \end{pmatrix}$

Hence, the single-qubit tomography experiment is straightforward. For an arbitrary one-qubit state, measurements are performed in three canonical bases. Photocounts are used to determine probabilities, which are employed to calculate Stokes parameters. The Stokes parameters finally fetch us the required density matrix. Table 6.1 also shows that the estimated density matrices correlate well with the theoretically predicted ones. We can now generalize this method for two-qubit tomography.

6.3 Two-qubit tomography

It is reasonably straightforward to generalize the tomography scheme based on the Stokes parameters to measure multi-qubit states. However, it must be kept in mind that there exist important differences between one-photon and two-photon cases. For instance, one-photon beams demonstrate polarization properties similar to those of classical optical beams and hence can be described in a purely classical manner [19], if differences in photodetection statistics are ignored. For two-photon or bi-partite systems, however, nonclassical correlations may also exist between the two or more beams. This is due to the unusual, purely quantum property of quantum entanglement.

In this section, we are interested in two-photon states. These states live in a four-dimensional space and are represented by a 4×1 column vector (if pure). For example, an arbitrary two-photon pure state can be written as

$$|\psi\rangle = a_0|HH\rangle + a_1|HV\rangle + a_2|VH\rangle + a_3|VV\rangle = \begin{pmatrix} a_0 \\ a_1 \\ a_2 \\ a_3 \end{pmatrix} \tag{6.26}$$

with complex numbers a_i and $\sum_{i=0}^{3}|a_i|^2 = 1$. The generalized mixed state can be written as the density matrix

$$\hat{\rho} = \begin{pmatrix} A_1 & B_1e^{i\phi_1} & B_2e^{i\phi_2} & B_3e^{i\phi_3} \\ B_1e^{-i\phi_1} & A_2 & B_4e^{i\phi_4} & B_5e^{i\phi_5} \\ B_2e^{-i\phi_2} & B_4e^{-i\phi_4} & A_3 & B_6e^{i\phi_6} \\ B_3e^{-i\phi_3} & B_5e^{-i\phi_5} & B_6e^{-i\phi_6} & A_4 \end{pmatrix}, \tag{6.27}$$

which makes $\hat{\rho}$ semi-definite positive, Hermitian and gives a trace equal to 1 $(A_1 + A_2 + A_3 + A_4 = 1)$. Hence, we need $16 - 1 = 15$ parameters to completely determine this matrix. These parameters are $(A_1, \cdots, A_4, B_1, \cdots, B_6, \phi_1, \cdots, \phi_6)$. For this two-qubit state, the equation analogous to equation (6.15) is

$$\hat{\rho} = \frac{1}{4}\sum_{i,j=0}^{3} S_{ij}\hat{\sigma}_i \otimes \hat{\sigma}_j, \tag{6.28}$$

and the S-coefficients are generally given as

$$S_{ij} = S_i \otimes S_j = (P_{|\psi_i\rangle} \pm P_{|\psi_i\perp\rangle}) \otimes (P_{|\psi_j\rangle} \pm P_{|\psi_j\perp\rangle}). \tag{6.29}$$

In this notation, $P_{|\psi_i\rangle}$ is the probability of projecting a single qubit state to the output channel designated as $|\psi_i\rangle$ in a beam splitter experiment that differentiates between $|\psi_i\rangle$ and its orthogonal state $|\psi_i\perp\rangle$. For normalization, it is required that $S_{00} = 1$. Hence, 15 real parameters are required to identify any point in the two-qubit, four-dimensional Hilbert space. These are in fact two-qubit versions of the Stokes parameters for single-qubit states. However, unlike the Bloch sphere for one-qubit states, we do not have an accessible graphical picture of this two-qubit space. Yet, the Stokes parameters and the measurement probabilities are still related in the two-qubit space [2, 20].

The two-qubit S-coefficients can all be spelled out also showing the meaning of the notation for the tensor product $\otimes$ used in equation (6.29),

$$
\begin{aligned}
S_{00} &= (P_{|H\rangle} + P_{|V\rangle}) \otimes (P_{|H\rangle} + P_{|V\rangle}) = P_{|HH\rangle} + P_{|HV\rangle} + P_{|VH\rangle} + P_{|VV\rangle} \\
S_{01} &= (P_{|H\rangle} + P_{|V\rangle}) \otimes (P_{|D\rangle} - P_{|A\rangle}) = P_{|HD\rangle} - P_{|HA\rangle} + P_{|VD\rangle} - P_{|VA\rangle} \\
S_{02} &= (P_{|H\rangle} + P_{|V\rangle}) \otimes (P_{|L\rangle} - P_{|R\rangle}) = P_{|HL\rangle} - P_{|HR\rangle} + P_{|VL\rangle} - P_{|VR\rangle} \\
S_{03} &= (P_{|H\rangle} + P_{|V\rangle}) \otimes (P_{|H\rangle} - P_{|V\rangle}) = P_{|HH\rangle} - P_{|HV\rangle} + P_{|VH\rangle} - P_{|VV\rangle} \\
S_{10} &= (P_{|D\rangle} - P_{|A\rangle}) \otimes (P_{|H\rangle} + P_{|V\rangle}) = P_{|DH\rangle} + P_{|DV\rangle} - P_{|AH\rangle} - P_{|AV\rangle} \\
S_{11} &= (P_{|D\rangle} - P_{|A\rangle}) \otimes (P_{|D\rangle} - P_{|A\rangle}) = P_{|DD\rangle} - P_{|DA\rangle} - P_{|AD\rangle} + P_{|AA\rangle} \\
S_{12} &= (P_{|D\rangle} - P_{|A\rangle}) \otimes (P_{|L\rangle} - P_{|R\rangle}) = P_{|DL\rangle} - P_{|DR\rangle} - P_{|AL\rangle} + P_{|AR\rangle} \\
S_{13} &= (P_{|D\rangle} - P_{|A\rangle}) \otimes (P_{|H\rangle} - P_{|V\rangle}) = P_{|DH\rangle} - P_{|DV\rangle} - P_{|AH\rangle} + P_{|AV\rangle} \\
S_{20} &= (P_{|L\rangle} - P_{|R\rangle}) \otimes (P_{|H\rangle} + P_{|V\rangle}) = P_{|LH\rangle} + P_{|LV\rangle} - P_{|RH\rangle} - P_{|RV\rangle} \\
S_{21} &= (P_{|L\rangle} - P_{|R\rangle}) \otimes (P_{|D\rangle} - P_{|A\rangle}) = P_{|LD\rangle} - P_{|LA\rangle} - P_{|RD\rangle} + P_{|RA\rangle} \\
S_{22} &= (P_{|L\rangle} - P_{|R\rangle}) \otimes (P_{|L\rangle} - P_{|R\rangle}) = P_{|LL\rangle} - P_{|LR\rangle} - P_{|RL\rangle} + P_{|RR\rangle} \\
S_{23} &= (P_{|L\rangle} - P_{|R\rangle}) \otimes (P_{|H\rangle} - P_{|V\rangle}) = P_{|LH\rangle} - P_{|LV\rangle} - P_{|RH\rangle} + P_{|RV\rangle} \\
S_{30} &= (P_{|H\rangle} - P_{|V\rangle}) \otimes (P_{|H\rangle} + P_{|V\rangle}) = P_{|HH\rangle} + P_{|HV\rangle} - P_{|VH\rangle} - P_{|VV\rangle} \\
S_{31} &= (P_{|H\rangle} - P_{|V\rangle}) \otimes (P_{|D\rangle} - P_{|A\rangle}) = P_{|HD\rangle} - P_{|HA\rangle} - P_{|VD\rangle} + P_{|VA\rangle} \\
S_{32} &= (P_{|H\rangle} - P_{|V\rangle}) \otimes (P_{|L\rangle} - P_{|R\rangle}) = P_{|HL\rangle} - P_{|HR\rangle} - P_{|VL\rangle} + P_{|VR\rangle} \\
S_{33} &= (P_{|H\rangle} - P_{|V\rangle}) \otimes (P_{|H\rangle} - P_{|V\rangle}) = P_{|HH\rangle} - P_{|HV\rangle} - P_{|VH\rangle} + P_{|VV\rangle}.
\end{aligned}
\tag{6.30}
$$

In these 16 formulations, the term $P_{|HD\rangle}$, for example, means the joint probability of detecting the first photon in the channel $|H\rangle$ when it is subject to an HV beam splitter and the second photon is detected in the channel $|D\rangle$ when this second photon is analyzed by an 'effective' DA beam splitter. All other definitions follow.

Now in order to estimate the density matrix, we need to determine all the S-coefficients. We can also upgrade the discussion on resource requirements for the single qubit to the two qubit case. Suppose we use two detectors for each qubit, a total of four. For each qubit, we can use three settings corresponding to the projections along the three bases: $\{|D\rangle, |A\rangle\}$, $\{|L\rangle, |R\rangle\}$ and $\{|H\rangle, |V\rangle\}$. This means that $3 \times 3 = 3^2 = 9$ settings are required. In each setting, we measure four ($2^2 = 4$) probabilities corresponding to the output channels $|HH\rangle$, $|HV\rangle$, $|VH\rangle$, $|VV\rangle$. This shows that $9 \times 4 = 36$ measurements are sufficient to provide information about the state $\hat{\rho}$. But there are only fifteen independent parameters, so clearly we have more measurements than are actually needed. We have an overdetermined system.

The density matrix can be used as a tool to characterize any state, determine the purity, degree of mixedness or quantify the amount of entanglement. For instance, *fidelity* is a quantity defined to measure the state overlap or 'likeness' of two states. For states represented by density matrices $\hat{\rho}_1$ and $\hat{\rho}_2$, fidelity F is generally given by [17]

$$F(\hat{\rho}_1, \hat{\rho}_2) = \left(\mathrm{Tr}\left[\sqrt{\sqrt{\hat{\rho}_1}\hat{\rho}_2\sqrt{\hat{\rho}_1}}\right]\right)^2. \tag{6.31}$$

Furthermore, if one of the two states being compared is pure, the fidelity expression simplifies to

$$F(\hat{\rho}_1, \hat{\rho}_2) = \mathrm{Tr}(\hat{\rho}_1\hat{\rho}_2). \tag{6.32}$$

Hence, if we want to check how well a particular entangled state is generated, we can do quantum state tomography of the generated state and check its fidelity against the theoretical prediction.

Two other figures of merit called *concurrence* and *tangle* also deserve attention. They closely related measures are used to quantify the entanglement of a system [8, 21, 22]. For a two-qubit system such as ours, concurrence C is defined as

$$C = \mathrm{Max}\left\{0,\ \sqrt{\lambda_1} - \sqrt{\lambda_2} - \sqrt{\lambda_3} - \sqrt{\lambda_4}\right\} \tag{6.33}$$

where $\lambda_1 > \lambda_2 > \lambda_3 > \lambda_4$ are the eigenvalues of the matrix given by $\hat{\rho}\hat{Z}\hat{\rho}^T\hat{Z}$. Superscript T denotes the transpose function and $\hat{Z}$ is called a 'spin flip matrix' defined as

$$\hat{Z} \equiv \begin{pmatrix} 0 & 0 & 0 & -1 \\ 0 & 0 & 1 & 0 \\ 0 & 1 & 0 & 0 \\ -1 & 0 & 0 & 0 \end{pmatrix}. \tag{6.34}$$

Tangle T is simply obtained from concurrence using the relation $T = C^2$. Both concurrence and tangle range from 0 for non-entangled or mixed states to 1 for maximally mixed states.

6.4 Nonideal measurements and compensation of errors

The technique described above empowers us to perfectly reconstruct the density matrix only if we have infinitely many ideal measurements. In real experiments, we cannot measure ideal probabilities and measurement is never perfect too. Moreover, imperfect wave plate orientations also result in measurements that are in bases which may be slightly different from the intended ones. On the other hand, a density matrix corresponding to any physical state must be positive semi-definite. Coupled with normalization and Hermiticity, it implies that all the eigenvalues of the density matrix must lie between 0 and 1 (inclusive), and that their sum must be 1. This, in turn, implies that $0 < \mathrm{Tr}(\hat{\rho}) < 1$. It is well known that these condition are not achieved in all experimental tomographic measurements [8, 23]

The errors propagating into the density matrix can be categorized into three types [8]. First of all, we can have errors in the measurement basis. These errors can be tackled by using an increasingly accurate measurement apparatus. Second, we have errors resulting from the counting statistics. Such errors can be reduced by taking a larger number of measurements, i.e. recording photocounts for a longer time. Third, we may have errors due to lack of experimental stability. These errors originate from variability of the generated state or the instability of the measurement apparatus. The variability in the detected intensity of the state (i.e. the rate of photons produced) is called drift. It can be somewhat compensated by using four detectors (instead of two) for the two-photon case [23]. The idea is that since each bi-partite member of an ensemble is measured in a complete basis, we do not need to assume a constant ensemble size for each measurement. For example, using a four-detector system with $\{|H\rangle, |V\rangle\}$ basis, we can obtain the photocounts N_{HH}, N_{HV}, N_{VH}, and N_{VV} for a finite time period. From these counts, total photocounts and probabilities can be determined confidently. If there is a drift in state intensity during a change in measurement basis, it will not harm the tomography results significantly since the total ensemble size is determined separately for each measurement.

As we discussed in chapter 4, there are accidental coincidence counts that are contributed by background light. Then there are detector's dark counts. These counts can all be subtracted from the total measured counts.

Even after taking care of all of these errors, it is possible that the state estimation procedure fetches us an illegal density matrix (e.g. a single-qubit state with radius greater than 1 in the Bloch space). This problem is taken care of by finding instead the legal state which most likely might have returned the counts that have been measured [2, 3]. This is the essence of the maximum likelihood technique which is utilized in the QST experiments briefly explained in the next section. Maximum likelihood estimation finds the parameters which may have made the outcomes most likely, provided a certain model is assumed. Our unknown parameters are the elements of the density matrix $\hat{\rho}$ which are given by $\{A_1, A_2, A_3, A_4, B_1, B_2, B_3, B_4, B_5, B_6, \phi_1, \phi_2, \phi_3, \phi_4, \phi_5, \phi_6\}$, the experimental outcomes are probabilities and the model is the set of equations (6.28)–(6.30).

6.5 Maximum likelihood estimation

The maximum likelihood technique provides an effective way to accommodate for imperfect measurements [8]. This technique resolves the issue of illegal density matrices by finding the state that is most likely to have resulted in the measured photocounts [2, 3]. Determining this legitimate state analytically is an extremely nontrivial task, and therefore we resort to numerical optimization. Maximum likelihood estimation requires three major elements: expression of a general density matrix in terms of legitimate parameters, a likelihood function that can be maximized, and a numerical technique for performing this optimization. The numerical technique must perform this maximization over the space of density matrix parameters.

Here, a legitimate state implies one having a non-negative definite Hermitian density matrix with a unit trace. The density matrix is expressed in the form of an accompanying matrix $\hat{W}$ such that

$$\hat{\rho}_p = \frac{\hat{W}^\dagger \hat{W}}{\mathrm{Tr}\{\hat{W}^\dagger \hat{W}\}} \tag{6.35}$$

fulfills the aforementioned requirements for valid density matrices for physical systems [2]. The subscript p in $\hat{\rho}_p$ signifies a physically legitimate density matrix. Dealing with the two-qubit system, it is a 4×4 density matrix with 15 independent real parameters. It is convenient to choose a tri-diagonal form for $\hat{W}$ [2] given by

$$\hat{W} = \begin{pmatrix} w_1 & 0 & 0 & 0 \\ w_5 + iw_6 & w_2 & 0 & 0 \\ w_7 + iw_8 & w_9 + iw_{10} & w_3 & 0 \\ w_{11} + iw_{12} & w_{13} + iw_{14} & w_{15} + iw_{16} & w_4 \end{pmatrix}. \tag{6.36}$$

Consider the measurement data consisting of a set of 36 coincidence counts n_i whose expected values are given by $\bar{n}_i = N\langle\psi_i|\hat{\rho}_p|\psi_i\rangle$ (for $i = 1, 2, \ldots, 36$), where N is a normalization parameter equal to the size of the ensemble per measurement. Then, assuming a Gaussian probability distribution of the coincidence counts, the optimization problem can be stated as being equivalent to finding the minimum of the function [2]

$$\mathcal{L} = \sum_{i=1}^{36} \frac{[N\langle\psi_i|\hat{\rho}_p|\psi_i\rangle - n_i]^2}{2N\langle\psi_i|\hat{\rho}_p|\psi_i\rangle}, \tag{6.37}$$

where n_i is the result of ith measurement. The function $\mathscr{L}$ is called the log-likelihood function. The final step in the method is an optimization routine. We use the MATLAB toolbox developed by the Kwiat group[2] for the aforementioned maximum likelihood estimation [8], and discuss the two-qubit experiment in the next section.

6.6 The experiment

We use a four-detector system for two qubits. The experimental setup is an extension to that of experiments NL2–NL3. Only two quarter-wave plates (QWPs) are added in the paths of the two downconverted beams. A schematic diagram of the setup is shown in figure 6.5 whereas a photograph is shown in figure 6.6. This is the extension of the single qubit case illustrated in figure 6.4(b) to two qubits.

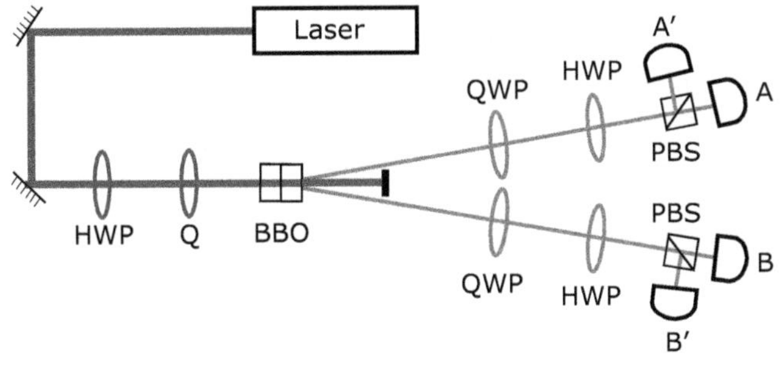

Figure 6.5. Schematic diagram of the two-qubit tomography experiment. In each path of the downconverted photon pair, a quarter-wave plate (QWP), a half-wave plate (HWP), a polarizing beam splitter (PBS) and two detectors are placed. The pump beam half-wave plate (HWP) and quartz plate (Q) are used to generate the required input state.

[2] http://research.physics.illinois.edu/QI/Photonics/tomography/.

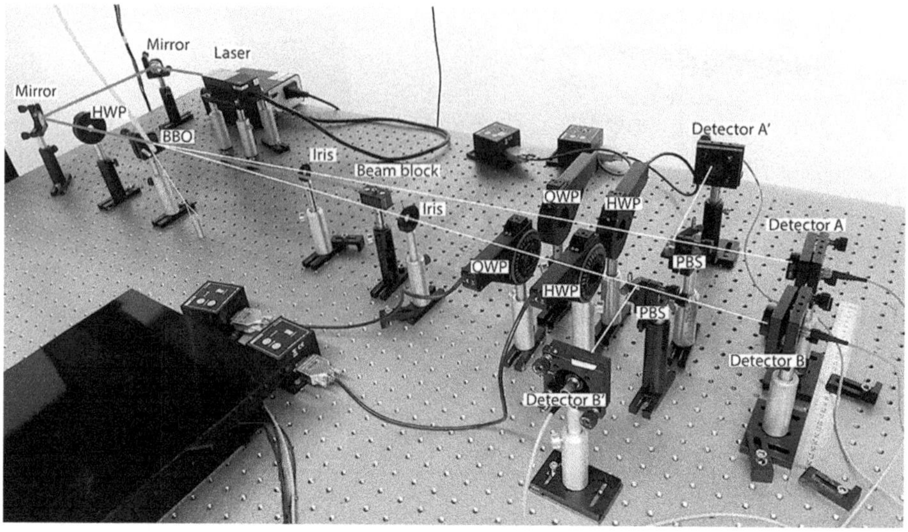

Figure 6.6. Photograph of the two-qubit tomography setup, excluding the quartz plate.

As shown in experiment Q3, we can perform an arbitrary polarization measurement and its orthogonal complement using a QWP, a half-wave plate (HWP) and a polarizing beam splitter (PBS). During the tomography experiment, we only change the orientation of the waveplates using motorized mounts to realize measurements in different polarization bases, whereas the input states are modified by changing the orientation of the pump beam HWP and rotation of the quartz plate with respect to the vertical axis.

Consider the general state (5.1) with parameters A, B and ϕ. The coefficients A and B are controlled by the orientation of the HWP. The phase ϕ can be effectively modified by changing the rotation of the quartz plate, which is basically a birefringent plate that adds a tunable phase factor to the $|VV\rangle$ term. We have already described the use of this phase-correcting element in chapter 5, sections 5.3 and 5.4.3. In this way, we can generate polarization states that are arbitrary linear combinations of the $|HH\rangle$ and $|VV\rangle$ states. For the $|HH\rangle$ and $|VV\rangle$ states, the quartz plate orientation does not matter. However, for the Bell state, the quartz plate is needed to eliminate the effective phase between the $|HH\rangle$ and $|VV\rangle$ terms.

For tomography of each generated state, we performed measurements in the nine settings, which along with the respective wave plate orientations are tabulated in table 6.2. Here we use the notations h_A, q_A, h_B and q_B for the half-wave and quarter-wave plate orientations for Alice (A) and Bob (B). All orientations are measured with respect to the lab's horizontal axis.

We perform tomography for three kinds of generated states $|HH\rangle$, $|VV\rangle$ and $(|HH\rangle + |VV\rangle)/\sqrt{2}$, using Kwiat's maximum likelihood estimation toolbox, compare the estimated density matrices with the theoretically predicted ones and compute the fidelity for each generated state (table 6.6). The datasets for the three states are presented in tables 6.3–6.5, respectively. Finally, the estimated density matrices are also plotted in figure 6.7.

Table 6.2. Measurement bases for two-qubit tomography. h_A (h_B) and q_A (q_B) represent the HWP and QWP orientations for Alice (Bob).

Measurement basis	h_A	q_A	h_B	q_B
$\|HH\rangle$	0°	0°	0°	0°
$\|HD\rangle$	0°	0°	22.5°	45°
$\|HL\rangle$	0°	0°	0°	45°
$\|DH\rangle$	22.5°	45°	0°	0°
$\|DD\rangle$	22.5°	45°	22.5°	45°
$\|DL\rangle$	22.5°	45°	0°	45°
$\|LH\rangle$	0°	45°	0°	0°
$\|LD\rangle$	0°	45°	22.5°	45°
$\|LL\rangle$	0°	45°	0°	45°

Table 6.3. Measurements for two-qubit tomography of the $|HH\rangle$ state. The numbers represent average measured coincidence counts per second. Accidental coincidence counts have been subtracted.

Measurement basis	AB	AB'	$A'B$	$A'B'$
$\|HH\rangle$	1691	41	1	0
$\|HD\rangle$	695	1124	0	7
$\|HL\rangle$	1414	441	0	6
$\|DH\rangle$	1004	20	791	0
$\|DD\rangle$	405	620	339	496
$\|DL\rangle$	857	196	302	208
$\|LH\rangle$	852	5	203	0
$\|LD\rangle$	322	165	72	449
$\|LL\rangle$	666	54	90	99

Table 6.4. Measurements for two-qubit tomography of the $|VV\rangle$ state. The numbers represent average measured coincidence counts per second. Accidental coincidence counts have been subtracted.

Measurement basis	AB	AB'	$A'B$	$A'B'$
$\|HH\rangle$	44	3	103	2031
$\|HD\rangle$	24	10	958	808
$\|HL\rangle$	47	2	1016	1583
$\|DH\rangle$	13	719	73	1173
$\|DD\rangle$	311	278	509	458
$\|DL\rangle$	155	557	412	1029
$\|LH\rangle$	52	1287	80	1365
$\|LD\rangle$	508	445	847	662
$\|LL\rangle$	146	961	874	1003

Table 6.5. Measurements for two-qubit tomography of the Bell state. The numbers represent average measured coincidence counts per second. Accidental coincidence counts have been subtracted.

Measurement Basis	AB	AB'	$A'B$	$A'B'$
$\|HH\rangle$	836	50	18	896
$\|HD\rangle$	433	615	297	437
$\|HL\rangle$	520	729	50	652
$\|DH\rangle$	566	197	371	459
$\|DD\rangle$	715	145	91	766
$\|DL\rangle$	354	596	371	242
$\|LH\rangle$	215	284	630	111
$\|LD\rangle$	332	621	324	360
$\|LL\rangle$	84	733	641	287

Table 6.6. Results of two-qubit quantum state tomography. For the estimated density matrices, only the real components of the matrices are shown.

State	Predicted density matrix	Re{Measured density matrix}	Fidelity
$\|HH\rangle$	$\begin{pmatrix} 1 & 0 & 0 & 0 \\ 0 & 0 & 0 & 0 \\ 0 & 0 & 0 & 0 \\ 0 & 0 & 0 & 0 \end{pmatrix}$	$\begin{pmatrix} 0.92 & -0.09 & 0.05 & -0.04 \\ -0.09 & 0.05 & 0.02 & 0.02 \\ 0.05 & 0.02 & 0.02 & 0.01 \\ -0.04 & 0.02 & 0.01 & 0.01 \end{pmatrix}$	0.92
$\|VV\rangle$	$\begin{pmatrix} 0 & 0 & 0 & 0 \\ 0 & 0 & 0 & 0 \\ 0 & 0 & 0 & 0 \\ 0 & 0 & 0 & 1 \end{pmatrix}$	$\begin{pmatrix} 0.01 & 0.00 & -0.01 & 0.06 \\ 0.00 & 0.02 & 0.00 & -0.10 \\ -0.01 & 0.00 & 0.05 & 0.04 \\ 0.06 & -0.10 & 0.04 & 0.92 \end{pmatrix}$	0.92
$\frac{\|HH\rangle + \|VV\rangle}{\sqrt{2}}$	$\begin{pmatrix} 0.5 & 0 & 0 & 0.5 \\ 0 & 0 & 0 & 0 \\ 0 & 0 & 0 & 0 \\ 0.5 & 0 & 0 & 0.5 \end{pmatrix}$	$\begin{pmatrix} 0.38 & -0.01 & -0.05 & 0.30 \\ -0.01 & 0.05 & 0.00 & 0.06 \\ -0.05 & 0.00 & 0.02 & -0.05 \\ 0.30 & 0.06 & -0.05 & 0.55 \end{pmatrix}$	0.77

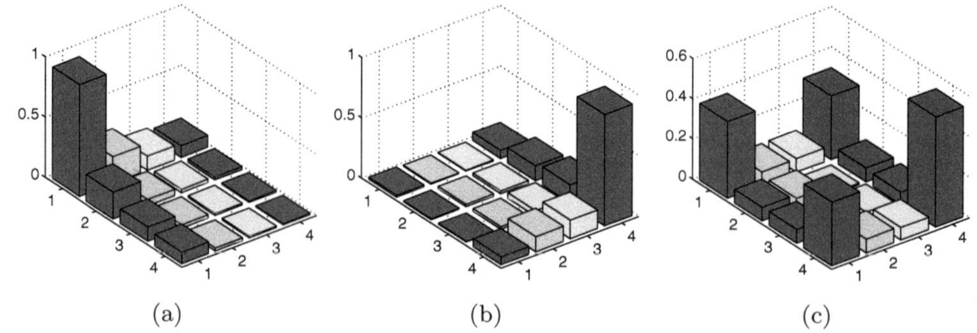

Figure 6.7. Experimentally estimated density matrices for (a) $|HH\rangle$, (b) $|VV\rangle$ and (c) $(|HH\rangle + |VV\rangle)/\sqrt{2}$ states. The plots are made using the absolute value of the respective density matrix components.

As discussed earlier, we can determine the purity of a state from its density matrix. For example, let's take the estimated density matrix for the Bell state $(|HH\rangle + |VV\rangle)/\sqrt{2}$. Its purity $\mathrm{Tr}(\hat{\rho}^2)$ is calculated to be 0.68. In contrast, if we take the predicted matrix which we also calculated in equation (3.31) in chapter 3, its purity comes out to be 1. For the generated Bell state, we also calculate concurrence (=0.61) and tangle (=0.37) which unsurprisingly are not equal to the ideal value (=1). The take-home message is that we desired to generate a pure, entangled state but due to nonideal experimental conditions we obtained an impure, mixed state.

Analyzing errors in the experimentally reconstructed density matrices is not a straightforward task. We do not perform error analysis here but briefly mention two methods. The traditional method analytically accounts for the errors in the measurements due to each identified source of error. The measurement errors are then propagated through calculations of derived quantities [2]. In this way, errors in the density matrices and its derived quantities, due to counting statistics and imperfect waveplates, are analyzed. Another method uses the Monte Carlo technique [8]. This technique revolves around using numerically simulated data to obtain additional density matrices, which are then utilized to compute errors in the quantities pertaining to the estimated density matrix.

References

[1] Stokes G G 1851 *Trans. Camb. Philos. Soc.* **9** 399

[2] James D F, Kwiat P G, Munro W J and White A G 2005 On the measurement of qubits *Asymptotic Theory of Quantum Statistical Inference: Selected Papers* (Singapore: World Scientific) pp 509–38

[3] Hradil Z 1997 *Phys. Rev.* A **55** R1561

[4] Tan S M 1997 *J. Mod. Opt.* **44** 2233

[5] Banaszek K, D'ariano G, Paris M and Sacchi M 1999 *Phys. Rev.* A **61** 010304

[6] Hradil Z, Summhammer J, Badurek G and Rauch H 2000 *Phys. Rev.* A **62** 014101

[7] Řeháček J, Hradil Z and Ježek M 2001 *Phys. Rev.* A **63** 040303

[8] Altepeter J B, Jeffrey E R and Kwiat P G 2005 *Adv. At. Mol. Opt. Phys.* **52** 105

[9] Ashburn J, Cline R, van der Burgt P J, Westerveld W and Risley J 1990 *Phys. Rev.* A **41** 2407

[10] Smithey D, Beck M, Raymer M G and Faridani A 1993 *Phys. Rev. Lett.* **70** 1244

[11] Dunn T, Walmsley I and Mukamel S 1995 *Phys. Rev. Lett.* **74** 884

[12] Leibfried D *et al* 1996 *Phys. Rev. Lett.* **77** 4281

[13] Kurtsiefer C, Pfau T and Mlynek J 1997 *Nature* **386** 150

[14] Klose G, Smith G and Jessen P S 2001 *Phys. Rev. Lett.* **86** 4721

[15] Chuang I L, Gershenfeld N and Kubinec M 1998 *Phys. Rev. Lett.* **80** 3408

[16] White A G, James D F, Eberhard P H and Kwiat P G 1999 *Phys. Rev. Lett.* **83** 3103

[17] Nielsen M A and Chuang I 2010 *Quantum Computation and Quantum Information* (Cambridge: Cambridge University Press)

[18] Hecht E 1998 *Optics* (Reading, MA: Addison-Wesley)

[19] Mandel L and Wolf E 1995 *Optical Coherence and Quantum Optics* (Cambridge: Cambridge University Press)

[20] Abouraddy A F, Sergienko A V, Saleh B E and Teich M C 2002 *Opt. Commun.* **201** 93
[21] Wootters W K 1998 *Phys. Rev. Lett.* **80** 2245
[22] Coffman V, Kundu J and Wootters W K 2000 *Phys. Rev.* A **61** 052306
[23] Altepeter J B, James D F and Kwiat P G 2004 *4 Qubit Quantum State Tomography Quantum State Estimation* (Berlin: Springer) pp 113–45

Chapter 7

Conclusion

Nowadays, many experiments are possible in the instructional physics laboratory using single photons. These experiments are modern, cost-effective versions of some of the most groundbreaking experiments that shaped our understanding of quantum mechanics in the previous century. Some of these experiments include demonstrating the existence of single photons [1], tests of Bell inequalities [2–4], single photon interference [5], quantum erasure [5], and quantum state tomography [6].

As narrated in this book, which might be used as a textbook for the laboratory, we have developed all the aforementioned experiments in our laboratory using a modular approach. In other words, subsequent experiments build on the ones appearing earlier in the book. This approach makes troubleshooting easier and boosts confidence for the experimenter as the simpler experiments are set up first.

We would like to comment on the relative difficulty of setting up the various experiments. As far as alignment is concerned, experiment Q1 requires two detectors and is the easiest one. Experiment Q4 can be directly built on it with addition of few optical elements. Experiment NL1 also follows from experiment Q1 but the entangled state generation makes the experiment a bit difficult. Next in terms of complexity level, come experiments Q2 and Q3, both of which require three detectors. However, once experiment Q2 is set up, experiment Q3 needs very little effort. The next difficult experiments include those involving four detectors and entanglement—experiments NL2, NL3 and QST. Finally, we found experiment Q5, incorporating single photon interference and quantum erasure, to be the most difficult. Setting up the interferometer and changing the interferometer path difference with the optimal finesse were the major challenges.

It goes without saying that many versions of these experiments have been developed and performed earlier in a number of colleges and universities, most of them in the United States [7–9]. We found the earlier discussions of these experiments very helpful in setting up our lab. To the best of our knowledge, ours is the first such lab in Pakistan and we are not aware of similar efforts in the region.

doi:10.1088/978-0-7503-3063-3ch7

As technology continues to advance, it is expected that more institutions will be able to set up their versions of quantum mechanical labs very soon. We hope that they will find this book a valuable resource.

We would like to conclude by mentioning a few of the wide-ranging benefits these experiments promise. They offer effective pedagogical tools to complement quantum mechanics courses. These experiments can motivate students to think about the fundamental aspects of quantum physics. This can also prove very beneficial for students who would like to investigate the foundations of quantum mechanics—an important area of research these days. Above all, these experiments can teach students a functional framework to think about and do research in quantum optics, quantum computing, quantum communication and quantum information.

References

[1] Thorn J *et al* 2004 *Am. J. Phys.* **72** 1210

[2] Dehlinger D and Mitchell M 2002 *Am. J. Phys.* **70** 903

[3] Carlson J, Olmstead M and Beck M 2006 *Am. J. Phys.* **74** 180

[4] Brody J and Selton C 2018 *Am. J. Phys.* **86** 412

[5] Beck M 2012 *Quantum Mechanics: Theory and Experiment* (Oxford: Oxford University Press)

[6] Altepeter J B, Jeffrey E R and Kwiat P G 2005 *Adv. At. Mol. Opt. Phys.* **52** 105

[7] Beck M and Galvez E J 2007 Quantum optics in the undergraduate teaching laboratory *Conf. on Coherence and Quantum Optics* (Washington, DC: Optical Society of America) p CSuA4

[8] Galvez E J 2019 Quantum optics laboratories for teaching quantum physics *Education and Training in Optics and Photonics* (Washington, DC: Optical Society of America) p 11143123

[9] Lukishova S G 2017 Quantum optics and nano-optics teaching laboratory for the undergraduate curriculum: teaching quantum mechanics and nano-physics with photon counting instrumentation *Education and Training in Optics and Photonics* (Washington, DC: Optical Society of America) p 104522I

IOP Publishing

Muhammad Hamza Waseem, Faizan-e-Ilahi and Muhammad Sabieh Anwar

Appendix A

Inventory for single photon experiments

Optical elements

Item	Quantity	Company	Part number
Laser-405 nm	1	CNI Laser	MDL-III-405-50 mW
Mirror	2	Thorlabs	PF05-03-P01
HWP-405 nm	1	Thorlabs	WPH05M-405
BBO	1	Newlight Photonics	PABBO5050-405(1)
HWP-808 nm	2	Thorlabs	WPH05M-808
QWP-808 nm	2	Thorlabs	WPQ05M-808
PBS	2	Thorlabs	PBS252
BDP	2	Thorlabs	BD40
Polarizer	2	Thorlabs	LPNIRE050-B
Iris	2	Thorlabs	ID25SS
Quartz crystal	1	MTI Corporation	SOX101005S2
Beam block	1	Thorlabs	LB1
Alignment Laser	1	Thorlabs	CPS180
Filters	4	Thorlabs	FGL780
Lens/collimators	5	Thorlabs	F220FC-780
Silica gel			

doi:10.1088/978-0-7503-3063-3ch8

Mechanical elements

Item	Quantity	Company	Part number
Optical table	1	Thorlabs	T1220CK
Kinematic mounts for mirrors	2	Thorlabs	KM100
Kinematic mounts for BBO	1	Thorlabs	KM100
Rotation mounts for wave plates	2	Thorlabs	RSP1(/M)
Motorized Rotation mounts for wave plates	4	Thorlabs	PRM1Z8
Rotation mounts for polarizer	2	Thorlabs	RSP05(/M)
Kinematic prism mounts for PBS	2	Thorlabs	KM100PM/M
Kinematic mounts for BDP	2	Thorlabs	KM100
Lens/Collimator adapter	5	Thorlabs	AD15F
Kinematic mounts for lens/collimator	5	Thorlabs	KC1/M
Fibre-fibre couplers	2	Thorlabs	FCB1
Clamping arm	3	Thorlabs	PM3SP/M
Post holder base		Thorlabs	BA1/M and BA2(/M)
Screws			M6
Posts		Homemade	
Post holders		Homemade	

Actuators and controllers

Item	Quantity	Company	Part number
DC servo motor controller for motorized rotation stage	4	Thorlabs	KDC101
Piezoelectric actuator for BDP mount	1	Thorlabs	PIA13
Piezo actuator controller	1	Thorlabs	KIM101

Detection and coincidence counting unit

Item	Quantity	Company	Part number
Detectors	4	Excelitas Technologies	SPCM-AQ4C
FPGA	1	Xilinx	Nexys 2
Power supplies	3	Extech	Extech 382213
Optical fibres	10	Thorlabs	M64L01
FPGA-APD connector (potential divider box)	1	Homemade	
BNC cables			
50 ohm BNC connectors			
Computer with LabVIEW	1		

Testing of FPGA

Item	Quantity	Company	Part number
FPGA	1	Xilinx	Nexys 2
Delay generator	1	Stanford Research Systems	DB64
Pulse generator	1	Stanford Research Systems	DG645
High speed oscilloscope	1	Agilent	DSO6104L
BNC cables			

The above tables incorporated the complete inventory for the lab. Below, we tabulate individual lists for each quantum experiment discussed in this book. In these lists, for the sake of brevity, we do not include obvious components such as optical table and computer or mounts and posts.

Q1: Spontaneous parametric downconversion

Item	Quantity	Company	Part number
Laser-405 nm	1	CNI Laser	MDL-III-405-50 mW
Mirror	2	Thorlabs	PF05-03-P01
HWP-405 nm	1	Thorlabs	WPH05M-405
BBO	1	Newlight Photonics	PABBO5050-405(1)
Beam block	1	Thorlabs	LB1
Iris	2	Thorlabs	ID25SS
Detector	2	Excelitas Technologies	SPCM-AQ4C

Q2: Proof of existence of photons

Item	Quantity	Company	Part number
Laser-405 nm	1	CNI Laser	MDL-III-405-50 mW
Mirror	2	Thorlabs	PF05-03-P01
HWP-405 nm	1	Thorlabs	WPH05M-405
BBO	1	Newlight Photonics	PABBO5050-405(1)
Beam block	1	Thorlabs	LB1
Iris	2	Thorlabs	ID25SS
HWP-808 nm	1	Thorlabs	WPH05M-808
PBS	1	Thorlabs	PBS252
Delay generator	1	Stanford Research Systems	DB64
Detector	3	Excelitas Technologies	SPCM-AQ4C

Q3: Estimating the polarization state of single photons

Item	Quantity	Company	Part number
Laser-405 nm	1	CNI Laser	MDL-III-405-50 mW
Mirror	2	Thorlabs	PF05-03-P01
HWP-405 nm	1	Thorlabs	WPH05M-405
BBO	1	Newlight Photonics	PABBO5050-405(1)
Beam block	1	Thorlabs	LB1
Iris	2	Thorlabs	ID25SS
HWP-808 nm	2	Thorlabs	WPH05M-808
QWP-808 nm	2	Thorlabs	WPQ05M-808
PBS	1	Thorlabs	PBS252
Detector	3	Excelitas Technologies	SPCM-AQ4C

Q4: Visualizing the polarization state of single photons

Item	Quantity	Company	Part number
Laser-405 nm	1	CNI Laser	MDL-III-405-50 mW
Mirror	2	Thorlabs	PF05-03-P01
HWP-405 nm	1	Thorlabs	WPH05M-405
BBO	1	Newlight Photonics	PABBO5050-405(1)
Beam block	1	Thorlabs	LB1
Iris	2	Thorlabs	ID25SS
HWP-808 nm	1	Thorlabs	WPH05M-808
QWP-808 nm	2	Thorlabs	WPQ05M-808
Polarizer	1	Thorlabs	LPNIRE050-B
Detector	2	Excelitas Technologies	SPCM-AQ4C

Q5: Single-photon interference and quantum eraser

Item	Quantity	Company	Part number
Laser-405 nm	1	CNI Laser	MDL-III-405-50 mW
Mirror	2	Thorlabs	PF05-03-P01
HWP-405 nm	1	Thorlabs	WPH05M-405
BBO	1	Newlight Photonics	PABBO5050-405(1)
Beam block	1	Thorlabs	LB1
Iris	2	Thorlabs	ID25SS
HWP-808 nm	2	Thorlabs	WPH05M-808
BDP	2	Thorlabs	BD40
Polarizer	1	Thorlabs	LPNIRE050-B
Detector	2	Excelitas Technologies	SPCM-AQ4C

NL1: Freedman's test of local realism

Item	Quantity	Company	Part number
Laser-405 nm	1	CNI Laser	MDL-III-405-50 mW
Mirror	2	Thorlabs	PF05-03-P01
HWP-405 nm	1	Thorlabs	WPH05M-405
Quartz crystal	1	MTI Corporation	SOX101005S2
BBO	1	Newlight Photonics	PABBO5050-405(1)
Beam block	1	Thorlabs	LB1
Iris	2	Thorlabs	ID25SS
Polarizer	2	Thorlabs	LPNIRE050-B
Detector	2	Excelitas Technologies	SPCM-AQ4C

NL2: Hardy's test of local realism

Item	Quantity	Company	Part number
Laser-405 nm	1	CNI Laser	MDL-III-405-50 mW
Mirror	2	Thorlabs	PF05-03-P01
HWP-405 nm	1	Thorlabs	WPH05M-405
Quartz crystal	1	MTI Corporation	SOX101005S2
BBO	1	Newlight Photonics	PABBO5050-405(1)
Beam block	1	Thorlabs	LB1
Iris	2	Thorlabs	ID25SS
HWP-808 nm	2	Thorlabs	WPH05M-808
PBS	2	Thorlabs	PBS252
Detector	4	Excelitas Technologies	SPCM-AQ4C

NL3: CHSH test of local realism

Item	Quantity	Company	Part number
Laser-405 nm	1	CNI Laser	MDL-III-405-50 mW
Mirror	2	Thorlabs	PF05-03-P01
HWP-405 nm	1	Thorlabs	WPH05M-405
Quartz crystal	1	MTI Corporation	SOX101005S2
BBO	1	Newlight Photonics	PABBO5050-405(1)
Beam block	1	Thorlabs	LB1
Iris	2	Thorlabs	ID25SS
HWP-808 nm	2	Thorlabs	WPH05M-808
PBS	2	Thorlabs	PBS252
Detector	4	Excelitas Technologies	SPCM-AQ4C

QST: Quantum state tomography

Item	Quantity	Company	Part number
Laser-405 nm	1	CNI Laser	MDL-III-405-50 mW
Mirror	2	Thorlabs	PF05-03-P01
HWP-405 nm	1	Thorlabs	WPH05M-405
Quartz crystal	1	MTI Corporation	SOX101005S2
BBO	1	Newlight Photonics	PABBO5050-405(1)
Beam block	1	Thorlabs	LB1
Iris	2	Thorlabs	ID25SS
QWP-808 nm	2	Thorlabs	WPQ05M-808
HWP-808 nm	2	Thorlabs	WPH05M-808
PBS	2	Thorlabs	PBS252
Detector	4	Excelitas Technologies	SPCM-AQ4C

Muhammad Hamza Waseem, Faizan-e-Ilahi and Muhammad Sabieh Anwar

Appendix B

Field-programmable gate array

Field-programmable gate arrays (abbreviated as FPGAs) are an essential part of modern electronics, particularly where flexible, high-performance functionality is desired. These days, they are an essential component of most instrumentation systems employed in physics experiments. The quantum experiments discussed in this book were based on an FPGA-based coincidence counting unit. This appendix is aimed at providing a gentle overview of the FPGA.

B.1 Introduction

An integrated circuit (IC) is made by integrating a large number of logic gates on a small chip of a semiconductor, usually silicon. Most ICs are application-specific and have a fixed functionality. Therefore, they are not well-suited to developing instrumentation where quick prototyping and newer functionalities are often needed. For such requirements, an FPGA is a good solution.

An FPGA is simply an IC containing a large number of re-programmable interconnected discrete logic elements. FPGAs are very useful for experimentalists because they allow quick and cost-effective development of new digital designs. They are reusable, flexible and can be used to make a vast variety of digital circuits. They offer a massively parallel architecture, processing many signals and logic operations simultaneously.

FPGAs are typically used for control and synchronization, data acquisition, digital signal processing, computer interfacing etc. To physicists working in particle physics and quantum optics, FPGAs are especially useful for trigger-based and coincidence counting experiments. Hence, understanding the FPGA is useful for designing and troubleshooting experiments.

FPGAs can be programmed even after being placed into a bigger circuit. In other words, the surrounding electronics can be operated as normal, while the FPGA is programmed. For educational and research purposes, they are available as part of circuit boards or development kits. For our quantum experiments detailed in this book, we use the Nexys 2 board developed by Xilinx.

doi:10.1088/978-0-7503-3063-3ch9

B.2 Architecture

An FPGA mainly comprises programmable logic blocks (which implement the logic operations), programmable inter-connects (wires and switches, which serve as connections between inputs/outputs (I/O) and logic devices) and programmable I/O blocks (which are at the periphery of the FPGA and are used for external connections). A schematic diagram of a basic FPGA architecture is shown in figure B1. Most modern FPGAs also contain additional circuitry, like memory blocks, multipliers, etc.

The major component of each logic block is a look-up table (LUT), which is responsible for the desired logic operation. An LUT is basically a small RAM on which a truth table is loaded. The user can configure the LUT to execute the required logic operation. To exemplify, a two-bit LUT and the associated logic operation (AND gate in this case) are depicted in figure B2. Note that the modern FPGAs contain LUTS with a larger number of inputs.

One can connect the output of one LUT to the input of another LUT to perform a complex logic operation. One can also store the output in a register (also included in each LUT) to be used later. All the routing of the signals between the logic blocks and the I/O blocks is performed by the inter-connects, which form a switching matrix. The I/O blocks are responsible for connection of the internal FPGA signals to the external circuitry.

In short, the configured LUTs and registers, glued together and to the I/O blocks by the inter-connects, determine the logical function of the circuit. A trivial circuit mapped onto the FPGA architecture is shown in figure B3.

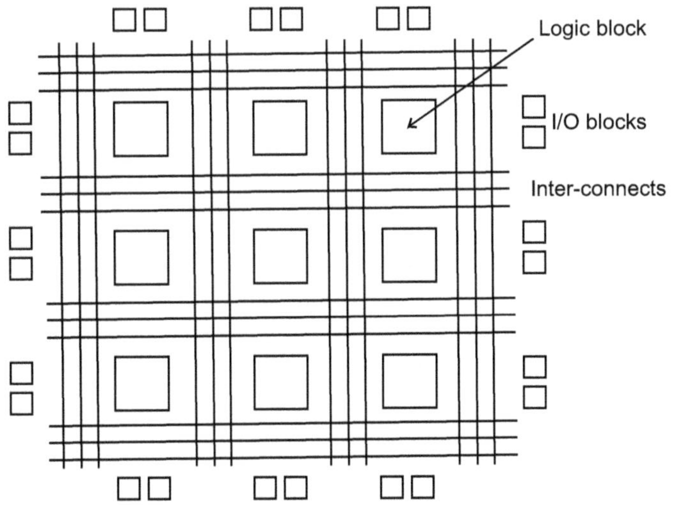

Figure B1. Basic architecture of an FPGA. Logic blocks interspersed in a matrix of inter-connects, with I/O blocks at the peripheries.

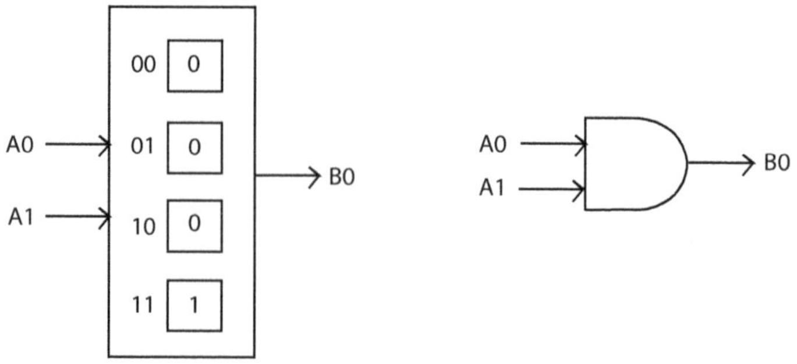

Figure B2. A two-bit look-up table (LUT) and the associated logic gate. If both the inputs, A0 and A1, are 1, the output B0 is 1. Otherwise, the output is zero. This LUT corresponds to an AND gate.

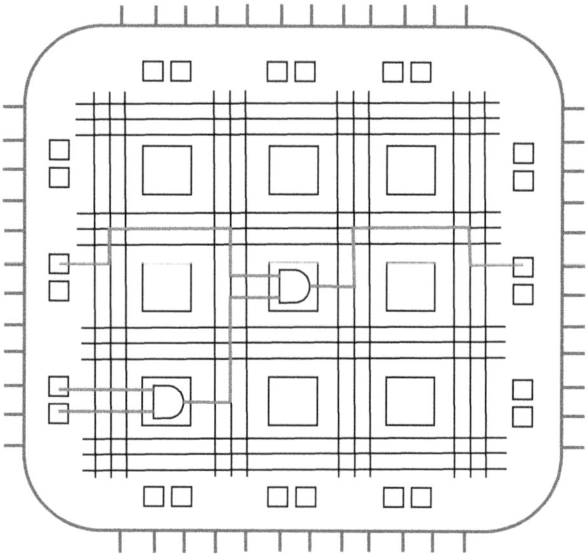

Figure B3. Configuration of an FPGA to perform a simple logic operation of ANDing three signals. The path in red shows the routing.

B.3 Configuring the FPGA

For configuring the FPGA, the user does not need to directly configure the bits of the logic blocks and the inter-connects. The FPGA manufactures offer development software or integrated development environment (IDE), which can compile a high-level description of the logic circuit and convert it into a configuration file, which is then downloaded onto the FPGA and configures the relevant bits. Let us look at the process in more detail.

B.3.1 Designing the system

Making a circuit design for an FPGA requires writing a description of the circuit's functionality using a hardware description language (HDL). The description or the code basically describes the logic operations as collection of processes which are to run parallel in the circuit. Most commonly used HDLs are Verilog and VHDL (we use Verilog for our FPGA). Alongside textual languages, one can also use graphical design tools to make circuit schematics. A synthesizer converts the HDL description of the system into a netlist, which is a description of the logic gates and the interconnections in the design. Simulations can be performed within the IDE to check the functionality of the design.

B.3.2 Implementing the design

Implementation of the design, step by step, as executed by the software, is as follows. A translator merges all the netlists as well as the design constraints. A mapper groups the gates in a netlist to efficiently utilize the LUTs of the FPGA. A place-and-route tool then places these grouped gates at different locations in the FPGA and also determines the routes between them. Finally, a bitstream is generated which contains a combination of 1s and 0s, encoding the routing settings for the interconnects and the truth-tables for the various LUTs.

This bitstream is the configuration file ready to be downloaded to the FPGA. When transferred to the FPGA, it configures the logical blocks and the connections between the blocks and the I/O blocks. In most cases, the FPGA is already soldered to a circuit board and once the configuration file is downloaded, the FPGA system is ready to be tested and used.

IOP Publishing

Muhammad Hamza Waseem, Faizan-e-Ilahi and Muhammad Sabieh Anwar

Appendix C

Coincidence counting unit code

This appendix lists a commented version of the complete program written for the coincidence counting unit. The code is written in Verilog, a common hardware description language, and is burnt on Xilinx Nexys 2 FPGA. It essentially counts pulses corresponding to single and coincidence photodetections, and transmits the count rates to a computer every 0.1 s.

Each quantum experiment discussed in this report uses the same program with minimal changes. To suit any particular experiment, only the coincidence signals need to be modified by making few changes in lines 42–45. The code is as follows.

```
// This is the main function which counts the single and
 coincidence photon detection pulses and sends the corresponding
 count rates to PC via UART every 1/10th of a second
module rs232coincidencecounter(output UART_TXD, input clock_50,
 input A, input B, input C, input D, output data_trigger, output
 baud_rate_clk, output wire Coincidence_0, output wire
 Coincidence_1, output wire Coincidence_2, output wire
 Coincidence_3);

// data_trigger is turned on every 1/10th of a second
 and begins the data stream out
// baud_rate_clk is the clock to output data at the baud
 rate of 19200 bits/second

// Counts the baud clock until it reaches 1920, which occurs
 every 1/10th of a second
  reg [14: 0] data_trigger_count;
```

doi:10.1088/978-0-7503-3063-3ch10

```
// Turns on every 1/10th of a second for one 50 MHz
  clock pulse signal and resets the photon detection counters
reg data_trigger_reset;

// Counts the 50 MHz clock pulses until it reaches 2604
  in order to time the baud clock
  reg [31: 0] baud_rate_count;

// Represents the top level design entity instantiation
  of the number of coincidences counted
  wire [31: 0] Count_top_0;
  wire [31: 0] Count_top_1;
  wire [31: 0] Count_top_2;
  wire [31: 0] Count_top_3;

// output registers of coincident photon counts
  reg [31: 0] Count_out_0;
  reg [31: 0] Count_out_1;
  reg [31: 0] Count_out_2;
  reg [31: 0] Count_out_3;

// Represents the top level design entity instantiation
  of the number of counts
  wire [31: 0] A_top;
  wire [31: 0] B_top;
  wire [31: 0] C_top;
  wire [31: 0] D_top;

// output registers of single photon counts
  reg [31: 0] A_out;
  reg [31: 0] B_out;
  reg [31: 0] C_out;
  reg [31: 0] D_out;

// Generation of four coincidence pulses from the input pulses
  coincidence_pulse CP0(.a(A), .b(B), .y(Coincidence_0));
  coincidence_pulse CP1(.a(A), .b(C), .y(Coincidence_1));
  coincidence_pulse CP3(.a(D), .b(B), .y(Coincidence_2));
  coincidence_pulse CP2(.a(D), .b(C), .y(Coincidence_3));

// Counts for a baud rate of 19200 and produces the baud rate clock signal
```

```
baud_rate_counter BRC1 (.clock_50(clock_50), .baud_rate_clk
  (baud_rate_clk));

// Uses the baud rate clock signal and generates a
  trigger signal every 1/10th of a second
  data_triggering
  DT1 (.baud_rate_clk(baud_rate_clk),
  .data_trigger(data_trigger));

// Outputs the data in 32 - bit registers and resets every 1/10th of a second
  counter C0 (.clock_50(clock_50), .data_trigger(data_trigger), .
pulse(Coincidence_0), .q(Count_top_0) );
  counter C1 (.clock_50(clock_50), .data_trigger(data_trigger), .
pulse(Coincidence_1), .q(Count_top_1) );
  counter C2 (.clock_50(clock_50), .data_trigger(data_trigger), .
pulse(Coincidence_2), .q(Count_top_2) );
  counter C3 (.clock_50(clock_50), .data_trigger(data_trigger), .
pulse(Coincidence_3), .q(Count_top_3) );
  counter CA (.clock_50(clock_50), .data_trigger(data_trigger), .
pulse(A), .q(A_top));
  counter CB (.clock_50(clock_50), .data_trigger(data_trigger), .
pulse(B), .q(B_top));
  counter CC (.clock_50(clock_50), .data_trigger(data_trigger), .
pulse(C), .q(C_top));
  counter CD (.clock_50(clock_50), .data_trigger(data_trigger), .
pulse(D), .q(D_top));

// This process updates the counts output arrays every 1/10th of a second
  always@(posedge data_trigger)
  begin
  A_out  <= A_top;
  B_out  <= B_top;
  C_out  <= C_top;
  D_out  <= D_top;
  Count_out_0  <= Count_top_0;
  Count_out_1  <= Count_top_1;
  Count_out_2  <= Count_top_2;
  Count_out_3  <= Count_top_3;
  end
```

```
// Sends the A, B, C, D
  and the coincidence counts out through the UART interface
  data_out D0 (.A(A_out), .B(B_out), .C(C_out), .D(D_out), .
coincidence_0(Count_out_0), .
coincidence_1(Count_out_1), .
coincidence_2(Count_out_2), .
coincidence_3(Count_out_3), .clk(baud_rate_clk),
  .data_trigger(data_trigger), .UART_TXD(UART_TXD));
  endmodule
// This function ANDs two pulse signals to form one coincidence
 pulse signal
module coincidence_pulse (input a, input b, output reg y);
  always @(*)
  begin
  y = a && b;
  end
  endmodule
// This function uses the baud rate clock signal and generates
 a trigger signal every 1/10th of a second
module data_triggering (input baud_rate_clk
  , output reg data_trigger);
  reg [31: 0] data_trigger_count;
   always @(posedge baud_rate_clk)
  begin
  data_trigger_count < = data_trigger_count + 1;
  if (data_trigger_count == 15'b000011110000000)
   begin
   data_trigger < = 1;
   data_trigger_count < =0;
   end
  else
   data_trigger < = 0;
  end
  endmodule
```

```
// This counter specifically counts for a baud rate of 19200 and
  produces a corresponding baud rate clock signal
module baud_rate_counter (input clock_50, output reg baud_rate_clk)
  ;
  reg [31: 0] baud_rate_count;
  always@(posedge clock_50)
   begin
   baud_rate_count <= baud_rate_count + 1;
   if (baud_rate_count >= 2604)
   begin
    baud_rate_clk <= 1;
    baud_rate_count <=0;
   end
    else
  baud_rate_clk <= 0;
  end
  endmodule
// This function counts voltage pulses
module counter(input clock_50, input data_trigger, input pulse,
  output reg [31: 0]q);
   wire x;
  or o1 (x, data_trigger, pulse);
  always @ (posedge x)
  begin
    if (data_trigger)
     q <=0;
    else
     q <=q + 1;
    end
  endmodule
// This function sends out up to four single photon counts and up
  to four coincidence counts to the PC through serial communication (UART)
module data_out(input [31: 0] A, input [31: 0] B, input [31: 0] C,
  input [31: 0] D, input [31: 0] coincidence_0, input [31: 0]
```

```
coincidence_1, input [31: 0] coincidence_2,
 input [31: 0] coincidence_3, input clk, input data_trigger,
 output reg
UART_TXD);
 reg [5: 0] index; begin
 reg [31: 0] out;
 reg [2: 0] data_select;
 always @ (posedge clk)
 begin
 if (index == 6'b111111 && data_trigger == 1)
  begin
  index < = 6'b000000;
  UART_TXD < = 1;
  out < = A;
  data_select < = 3'b000;
  end
 else if (index == 6'b000000)
  begin
  index < = 6'b000001;
  UART_TXD < = 0;
  end
   else if (index == 6'b000001)
   begin
   index < = 6'b000010;
   UART_TXD < = out[0];
   end
   else if (index == 6'b000010)
    begin
    index < = 6'b000011;
    UART_TXD < = out[1];
    end
   else if (index == 6'b000011)
    begin
    index < = 6'b000100;
    UART_TXD < = out[2];
    end
```

```
else if (index == 6'b000100)
 begin
 index <= 6'b000101;
 UART_TXD <= out[3];
 end
else if (index == 6'b000101)
 begin
 index <= 6'b000110;
 UART_TXD <= out[4];
 end
else if (index == 6'b000110)
 begin
 index <= 6'b000111;
 UART_TXD <= out[5];
 end
else if (index == 6'b000111)
 begin
 index <= 6'b001000;
 UART_TXD <= out[6];
 end
else if (index == 6'b001000)
 begin
 index <= 6'b001001;
 UART_TXD <= 0;
    end
else if (index == 6'b001001)
 begin
 index <= 6'b001010;
 UART_TXD <= 1; // the first stop bit
 end
else if (index == 6'b001010)
 begin
 index <= 6'b001011;
 UART_TXD <= 0; // the second start bit
 end
```

```
else if (index == 6'b001011)
 begin
 index <= 6'b001100;
 UART_TXD <= out[7];
 end
else if (index == 6'b001100)
 begin
 index <= 6'b001101;
 UART_TXD <= out[8];
 end
else if (index == 6'b001101)
 begin
 index <= 6'b001110;
 UART_TXD <= out[9];
 end
else if (index == 6'b001110)
 begin
 index <= 6'b001111;
 UART_TXD <= out[10];
 end
else if (index == 6'b001111)
 begin
 index <= 6'b010000;
 UART_TXD <= out[11];
 end
else if (index == 6'b010000)
begin
index <= 6'b010001;
UART_TXD <= out[12];
end
else if (index == 6'b010001)
 begin
 index <= 6'b010010;
 UART_TXD <= out[13];
 end
else if (index == 6'b010010)
```

```
begin
index <= 6'b010011;
UART_TXD <= 0; // the termination bit
end

else if (index == 6'b010011)
begin
index <= 6'b010100;
UART_TXD <= 1; // the second stop bit
end
else if (index == 6'b010100)
begin
index <= 6'b010101;
UART_TXD <= 0; // the third start bit
end
else if (index == 6'b010101)
begin
index <= 6'b010110;
UART_TXD <= out[14];
end

else if (index == 6'b010110)
begin
index <= 6'b010111;
UART_TXD <= out[15];
end
else if (index == 6'b010111)
begin
index <= 6'b011000;
UART_TXD <= out[16];
end
else if (index == 6'b011000)
begin
index <= 6'b011001;
UART_TXD <= out[17];
end
else if (index == 6'b011001)
```

```
 begin
 index <= 6'b011010;
 UART_TXD <= out[18];
 end
else if (index == 6'b011010)
 begin
 index <= 6'b011011;
 UART_TXD <= out[19];
 end

else if (index == 6'b011011)
 begin
 index <= 6'b011100;
 UART_TXD <= out[20];
 end
else if (index == 6'b011100)
 begin
 index <= 6'b011101;
 UART_TXD <= 0; // the termination bit
 end
else if (index == 6'b011101)
 begin
 index <= 6'b011110;
 UART_TXD <= 1; // the third stop bit
 end

else if (index == 6'b011110)
 begin
 index <= 6'b011111;
 UART_TXD <= 0; // the fourth start bit
 end
else if (index == 6'b011111)
 begin
 index <= 6'b100000;
 UART_TXD <= out[21];
 end
```

```
else if (index == 6'b100000)
 begin
 index <= 6'b100001;
 UART_TXD <= out[22];
 end
else if (index == 6'b100001)
 begin
 index <= 6'b100010;
 UART_TXD <= out[23];
 end
else if (index == 6'b100010)
 begin
 index <= 6'b100011;
 UART_TXD <= out[24];
 end
else if (index == 6'b100011)
 begin
 index <= 6'b100100;
 UART_TXD <= out[25];
 end
else if (index == 6'b100100)
 begin
 index <= 6'b100101;
 UART_TXD <= out[26];
 end
else if (index == 6'b100101)
 begin
 index <= 6'b100110;
 UART_TXD <= out[27];
 end
else if (index == 6'b100110)
 begin
 index <= 6'b100111;
 UART_TXD <= 0; // termination bit
 end
```

```
else if (index == 6'b100111)
 begin
 index <= 6'b101000;
 UART_TXD <= 1; // the fourth stop bit
 end
else if (index == 6'b101000)
 begin
 index <= 6'b101001;
 UART_TXD <= 0; // the fifth start bit
 end
else if (index == 6'b101001)
 begin
 index <= 6'b101010;
UART_TXD <= out[28];
end
else if (index == 6'b101010)
 begin
 index <= 6'b101011;
 UART_TXD <= out[29];
 end
else if (index == 6'b101011)
 begin
 index <= 6'b101100;
 UART_TXD <= out[30];
 end
else if (index == 6'b101100)
 begin
 index <= 6'b101101;
 UART_TXD <= out[31];
 end
else if (index == 6'b101101)
 begin
 index <= 6'b101110;
 UART_TXD <= 0;
 end
```

```
else if (index == 6'b101110)
 begin
 index <= 6'b101111;
 UART_TXD <= 0;
 end
else if (index == 6'b101111)
 begin
 index <= 6'b110000;
 UART_TXD <= 0;
 end
else if (index == 6'b110000)
 begin
 index <= 6'b110001;
 UART_TXD <= 0;
 end
else if (index == 6'b110001 && data_select == 3'b000)
 begin
 index <= 6'b000000;
 data_select <= 3'b001; // increments data_select to begin output of B
 out <= B;
 UART_TXD <= 1; // the fifth stop bit
 end
else if (index == 6'b110001 && data_select == 3'b001)
 begin
 index <= 6'b000000;
 data_select <= 3'b010; // increments data_select to begin output of C
 out <= C;
 UART_TXD <= 1; // the fifth stop bit
 end
else if (index == 6'b110001 && data_select == 3'b010)
 begin
 index <= 6'b000000;
 data_select <= 3'b011; // increments data_select to begin output of D
 out <= D;
 UART_TXD <= 1; // the fifth stop bit
 end
else if (index == 6'b110001 && data_select == 3'b011)
```

```
 begin
 index <= 6'b000000;
 data_select <= 3'b100; // increments data_
                          select to begin output of Coincidence_0
out <= coincidence_0;
UART_TXD <= 1; // the fifth stop bit
end
else if (index == 6'b110001 && data_select == 3'b100)
 begin
 index <= 6'b000000;
 data_select <= 3'b101; // increments data_
                          select to begin output of Coincidence_1
 out <= coincidence_1;
 UART_TXD <= 1; // the fifth stop bit
 end

else if (index == 6'b110001 && data_select == 3'b101)
 begin
 index <= 6'b000000;
 data_select <= 3'b110; // increments data_
                          select to begin output of Coincidence_2
 out <= coincidence_2;
 UART_TXD <= 1; // the fifth stop bit
 end
else if (index == 6'b110001 && data_select == 3'b110)
 begin
 index <= 6'b000000;
 data_select <= 3'b111; // increments data_
                          select to begin output of Coincidence_3
 out <= coincidence_3;
 UART_TXD <= 1; // the fifth stop bit
 end
else if (index == 6'b110001 && data_select == 3'b111)
 begin
 index <= 6'b110010;
 UART_TXD <= 1; // the fifth stop bit
```

```
 end
 else if (index == 6'b110010)
 begin
 index <= 6'b111111;
 UART_TXD <= 0; // the start bit of the termination byte
 end
else
 begin
 index <= 6'b111111;
 UART_TXD <= 1; // sets all subsequent bits to negative voltage
 end
end
endmodule
```

www.ingramcontent.com/pod-product-compliance
Ingram Content Group UK Ltd.
Pitfield, Milton Keynes, MK11 3LW, UK
UKHW052040301026
9729IPUK00019B/338

9 780750 330640